Island Environments in a Changing World

Islands represent unique opportunities to examine human interaction with the natural environment. They capture the human imagination as remote, vulnerable, and exotic, yet there is comparatively little understanding of their basic geology, geography, or the impact of island colonization by plants, animals, and humans. This detailed study of island environments focuses on nine island groups, including Hawai'i, New Zealand, Japan, and the British Isles, exploring their differing geology, geography, climate, and soils, as well as the varying effects of human actions. It illustrates the natural and anthropogenic disturbances common to island groups, all of which face an uncertain future clouded by extinctions of endemic flora and fauna, growing populations of invasive species, and burgeoning resident and tourist populations.

Examining the natural and human history of each island group from early settlement onwards, the book provides a critique of the concept of sustainable growth and offers realistic guidelines for future island management.

LAWRENCE R. WALKER is a professor of plant ecology at the University of Nevada, Las Vegas. His research focuses on the mechanisms that drive plant succession after cyclones and on volcanoes, landslides, glacial moraines, floodplains, dunes, mine tailings, and abandoned roads. Much of his work has been conducted on islands.

PETER BELLINGHAM is a research scientist and plant ecologist at Landcare Research, Lincoln, New Zealand. He has worked extensively in island ecosystems on the consequences of natural disturbances, such as cyclones, earthquakes, floods, and landslides, and on the interactions between these natural disturbances and invasions by non-native plants and animals.

Island
Environments
in a Changing
World

LAWRENCE R. WALKER
*University of Nevada, Las Vegas,
USA*

PETER BELLINGHAM
*Landcare Research, Lincoln,
New Zealand*

CAMBRIDGE UNIVERSITY PRESS
Cambridge, New York, Melbourne, Madrid, Cape Town,
Singapore, São Paulo, Delhi, Tokyo, Mexico City

Cambridge University Press
The Edinburgh Building, Cambridge CB2 8RU, UK

Published in the United States of America
by Cambridge University Press, New York

www.cambridge.org
Information on this title: www.cambridge.org/9780521519601

First published 2011

Printed in the United Kingdom at the University Press, Cambridge

A catalogue record for this publication is available from the British Library

Library of Congress Cataloguing in Publication data
Walker, Lawrence R.
  Island environments in a changing world / Lawrence R. Walker, Peter Bellingham.
    p. cm.
  ISBN 978-0-521-51960-1 (hardback)
  1. Islands. 2. Islands – Environmental aspects. 3. Island ecology.
  I. Bellingham, Peter. II. Title.
  GB471.W35 2011
  577.5′2–dc22   2010051718

ISBN 978-0-521-51960-1 Hardback
ISBN 978-0-521-73247-5 Paperback

Contents

Preface   *page* xi
Photo credits   xiv

1 Introduction to island environments and cultures   1
1.1 Stereotypes and realities   1
1.2 Special features of islands   3
    1.2.1 Isolation   4
    1.2.2 Finiteness   4
    1.2.3 Vulnerability   6
    1.2.4 Evolutionary experiments   7
1.3 Human impacts   8
    1.3.1 Conservation   10
    1.3.2 Invasive species and their management 11
    1.3.3 Restoration   13
1.4 Our island framework   14
1.5 Scope   17

2 The physical setting   21
2.1 Isolation and finiteness   21
2.2 Geology   21
    2.2.1 Oceanic islands   22
    2.2.2 Continental islands and land bridges 28
    2.2.3 Continental fragments   29
    2.2.4 Barrier islands   33
    2.2.5 Anthropogenic islands   33
2.3 Geography   35
    2.3.1 British Isles   39
    2.3.2 Iceland   40

2.3.3 Canary Islands   41
2.3.4 Puerto Rico   43
2.3.5 Jamaica   43
2.3.6 Hawai'i   45
2.3.7 Tonga   46
2.3.8 New Zealand   46
2.3.9 Japan   48
2.4 Climate   50
2.4.1 Topography   51
2.4.2 Oceanicity   54
2.4.3 Precipitation   54
2.4.4 Temperature   55
2.4.5 Climate measurement and integration   56
2.5 Soils   57
2.5.1 Geological substrate   58
2.5.2 Topographical influences   60
2.5.3 Climatic influences   61
2.5.4 Biological influences   61
2.6 Summary   62

3   Natural disturbances on islands   64
3.1 Disturbance characteristics   64
3.2 Volcanoes   67
3.2.1 Characteristics and examples   67
3.2.2 Ecological effects and responses   70
3.3 Earthquakes   72
3.3.1 Characteristics and examples   72
3.3.2 Ecological effects and responses   78
3.4 Erosion   79
3.4.1 Characteristics and examples   79
3.4.2 Ecological effects and responses   83
3.5 Land building   83
3.5.1 Characteristics and examples   83
3.5.2 Ecological effects and responses   87
3.6 Tropical cyclones   87
3.6.1 Characteristics and examples   87
3.6.2 Ecological effects and responses   94
3.7 Floods   96
3.7.1 Characteristics and examples   96
3.7.2 Ecological effects and responses   101

3.8 Tsunamis   102
   3.8.1 Characteristics and examples   102
   3.8.2 Ecological effects and responses   105
3.9 Droughts   106
   3.9.1 Characteristics and examples   106
   3.9.2 Ecological effects and responses   107
3.10 Fires   108
   3.10.1 Characteristics and examples   108
   3.10.2 Ecological effects and responses   109
3.11 Animal activities   110
3.12 Summary   111

4   The plants and animals of islands   116
4.1 Introduction   116
4.2 How islands gain their plants and animals   117
   4.2.1 Dispersal   118
   4.2.2 Past land connections   125
4.3 Evolution of new species   128
   4.3.1 Time   128
   4.3.2 Isolation   129
   4.3.3 Size and topography   129
   4.3.4 Biological features   132
4.4 Special features of plant and animal communities   135
   4.4.1 Species richness   135
   4.4.2 Diverse but related species   136
   4.4.3 Unusual life forms and behaviors   140
   4.4.4 Strong connections between land and sea   143
4.5 Extinction on islands   146
   4.5.1 Extinction as an evolutionary process   146
   4.5.2 Modern extinctions   147
4.6 Summary   149

**5   Human dispersal, colonization, and early
environmental impacts   152**
5.1 Introduction   152
5.2 Walking to Britain   152
5.3 Japan's first settlers   156
5.4 Settling Puerto Rico and Jamaica   159
5.5 The first Canary Islanders   162
5.6 Early Polynesia – the settlement of Tonga   164

5.7 Reaching the edges of Polynesia – the discovery and
     settlement of Hawai'i   168
5.8 Settling the largest islands of Polynesia – reaching
     Aotearoa (New Zealand)   172
5.9 Colonizing Iceland – the Norse outpost   175
5.10 Summary   179

6   **Intensifying human impacts on islands**   **182**
6.1 Introduction   182
6.2 Deforestation of the British Isles and their
     conversion to agriculture   183
6.3 Japan: a civilization founded on rice cultivation
     and forest management   192
6.4 Canary Islands: from self-sufficiency to trading post
     and cash crops   202
6.5 Puerto Rico and Jamaica: conquest, slavery, and crops
     for empires   207
6.6 Polynesian islands, European agriculture,
     and ecological transformation   213
     6.6.1 The colonial transformation of New Zealand's
            environment   215
     6.6.2 Hawai'i joins the global markets   221
     6.6.3 Tonga retains local control of land   228
6.7 Iceland finds a path forward after environmental
     degradation   229
6.8 Common trends   231

7   **Islands in the modern world, 1950–2000**   **235**
7.1 Introduction   235
7.2 State of the environment in 1950   236
7.3 Technology   240
     7.3.1 Fishing   240
     7.3.2 Agriculture   242
     7.3.3 Extractive industries   251
     7.3.4 Military activities   256
7.4 Population growth   259
     7.4.1 Urbanization   259
     7.4.2 Remittance cultures   266
7.5 Wealth and leisure   267
     7.5.1 Tourism   267
     7.5.2 Sport hunting   275

7.6 Invasive species   277
   7.6.1 Dispersal   277
   7.6.2 Novel biological communities   278
7.7 Responses   279
   7.7.1 Conservation   279
   7.7.2 Restoration   283
   7.7.3 Environmental limits imposed on humans
      by island ecosystems   289
7.8 State of the environment in 2000   290

8 The future of island ecosystems: remoteness lost   293
8.1 Introduction   293
8.2 Population pressure   293
8.3 Climate change   296
   8.3.1 Temperature   297
   8.3.2 Sea level and acidity   298
   8.3.3 Coral reefs   299
   8.3.4 Forests   300
   8.3.5 Tourism and the carbon cost of travel   301
8.4 Responses   301
   8.4.1 Local production and consumption   301
   8.4.2 Restoration   302
   8.4.3 Living with invasive species   302
   8.4.4 Urban futures   303
8.5 Broader implications for the island groups   304
   8.5.1 Application to other islands   304
   8.5.2 Practical lessons   305
8.6 Summary   305

Glossary   307
Index   312

The color plates will be found between pages 82 and 83

Preface

What makes islands so special to many of us? Is it their remoteness and the special effort we make to get to them? Is it their sharp land–water boundaries decorated by pretty beaches or dramatic, rocky shorelines? Is it their volcanoes or coral reefs? Is it their unusual and often rare plants and animals? Or perhaps it is their exotic cultures and unique languages? Whatever the attraction, tens of millions of people visit islands each year. As scientists, we too enjoy the scenic and cultural attractions and have been drawn to study the fascinating island plants and the animals that interact with them. However, we recognize that the massive popularity of islands and their burgeoning populations are changing island environments and that islands are particularly vulnerable to global climate changes, such as increases in air and water temperatures, sea level rise, and ocean acidification. We wrote this book to examine how humans have interacted over time with island environments from their earliest colonization until the present.

Our findings suggest that because of human colonization and development, island environments have lost the isolation that protected them as evolutionary experiments. With finite habitats, limited resources, and little protection from humans and the organisms humans bring with them, island plants and animals are vulnerable to extinction. Further, widespread introductions of non-native species have resulted in a global homogenization of plants and animals. In turn, populations of humans on islands are vulnerable because they rely on healthy environments for their well-being. The interaction between humans and the environment is therefore worth examining, not only for aesthetic or moral reasons, but so that we can plan how to manage, conserve, and restore natural island habitats and the services that they supply to humans.

We cover nine island groups where we have worked, six of which we have lived on, thereby adding personal perspectives to our narrative. These include two large island groups near continents where a long history of human habitation has altered the landscape (Japan and the British Isles), two medium-sized island groups colonized relatively recently by humans but also with human impacts (New Zealand and Iceland), and five smaller island groups with variable human histories and impacts (Hawai'i, Jamaica, Puerto Rico, Canary Islands, and Tonga). This collection of islands provides a broad range of physical, biological, and cultural aspects that we use to contrast our themes of isolation, finiteness, and vulnerability. How island environments adapt to human impacts concerns all of us, whether we actually visit islands or not. Islands serve as early warning signals from which we can examine human impacts on a small scale, learn lessons for application to mainland habitats, and attempt to reverse the unfortunate trends of environmental damage.

Lawrence thanks his wife, Elizabeth Powell, for all her support and his many colleagues who have influenced his island research. These include in particular Kristín Svavarsdóttir (Iceland); Nick Brokaw, Fred Landau, Jean Lodge, Joanne Sharpe, Joe Wunderle, and Jess Zimmerman (Puerto Rico); Rob Allen, Richard Bardgett, Duane Peltzer, David Wardle, and Susan Wiser (New Zealand); and Marian Chau, Don Drake, Klaus Mehltreter, and Aaron Shiels (Hawai'i). Lawrence also thanks the programs and institutions that have supported his island research including the United States Fulbright Foundation (Iceland); US National Science Foundation Long-Term Ecological Research Program (Puerto Rico); Landcare Research (New Zealand); and Stanford University and the Wilder Chair Program in the Botany Department of the University of Hawai'i at Mānoa (Hawai'i).

Peter thanks his wife, Lynda Burns, who has traveled to many islands with him and supported him during the writing of this book, and he thanks Michael and Sylvia for their forbearance. He thanks the colleagues who have influenced his island research: Edmund Tanner and John Healey (Jamaica, British Isles); Peter Grubb (Canary Islands, British Isles); Takashi Kohyama, Peter Matthews (Japan); Fred Landau (Puerto Rico); Graham Nugent, Don Drake, and Aaron Shiels (Hawai'i); Gerard Fitzgerald (Tonga); and many of his New Zealand colleagues including Rob Allen, Ewen Cameron, Joe Davis, Phil Lyver, Matt McGlone, Christa Mulder, John Ogden, Duane Peltzer, David Towns, David Wardle, Janet Wilmshurst, Susan Wiser, and Anthony Wright. Peter thanks the programs and institutions that supported his island research including the UK Natural Environment Research Council and the Royal Society (Jamaica), the US National Science Foundation (Puerto Rico), the Japan Society for the Promotion of Science (Japan), the New Zealand Ministry of Foreign Affairs and Trade (Tonga), the US Nature Conservancy and the Botany Department of the University of Hawai'i at Mānoa (Hawai'i), a Manaaki Whenua fellowship from Landcare Research (Jamaica, Canary Islands, British Isles), and the New Zealand Foundation for Research Science and Technology (New Zealand; Te Hiringa Tangata ki te Tai Timu ki te Tai Pari program).

We appreciate the able assistance of Paula Garrett (University of Nevada, Las Vegas) and James Barringer (Landcare Research, New Zealand) with graphic design. We are indebted to several critical readers whose comments helped us improve all or parts of the book. These readers include: Don Drake, José María Fernández-Palacios, Scott Fitzpatrick, Christopher Hamlin, Matt McGlone, Roger del Moral, John Parkes, Elizabeth Powell, Fred Swanson, Shiro Tsuyuzaki, Margery Walker, and Janet Wilmshurst. The Ruamāhua Islands Trust approved inclusion of some material. Finally, we appreciated the enthusiastic support for this project from the Cambridge University Press editorial staff, in particular Dominic Lewis.

The publisher has used its best endeavors to ensure that URLs for external websites referred to in this book are correct at the time of going to press. However, the publisher has no responsibility for the websites and can make no guarantee that a site will remain active or that the content is or will remain appropriate.

Photo credits

Peter Bellingham: Plates 2, 5, 9, 13, 14, 15, 16; Figures 1.1, 1.4a and b, 1.5, 1.9, 2.1, 2.6, 2.8, 2.16, 2.20, 2.23, 2.24a and b, 3.3, 3.4, 3.7, 3.10, 3.11, 3.12, 3.14, 3.15, 3.17, 3.18, 4.4, 4.5, 4.6, 4.8, 4.9, 4.10, 4.13, 5.2, 5.5, 6.1, 6.3, 6.5, 6.7, 6.8, 6.9, 6.10, 6.11, 6.12, 6.13a, 6.14, 6.15, 6.17, 6.21a, 7.2, 7.5, 7.6, 7.7, 7.9, 7.10, 7.11, 7.12, 7.13, 8.1

Don Drake: Front cover; Plates 8, 10, 11; Figures 2.2, 3.13, 4.12, 6.13b, 6.22, 6.23

Grant Hunter: Figure 6.16

Matthew Lurie: Figure 5.6

Peter McGregor: Figure 3.16

Elizabeth Powell: Back cover; Figures 1.2, 2.10

Aaron Shiels: Figures 5.3, 5.4, 5.7, 6.4, 7.17

Lawrence Walker: Plates 1, 3, 4, 7, 12; Figures 1.7, 1.8, 1.10, 2.3, 2.4, 2.25, 3.1, 3.2, 3.5, 3.6, 3.9, 3.19, 4.1, 4.2, 4.7, 5.8, 5.9, 5.10, 6.2, 6.18, 6.19, 6.20, 6.21b, 6.24, 7.3, 7.14, 7.15a and b, 7.16, 7.18a and b, 7.19, 8.2

Anthony Wright: Figure 4.14

1

Introduction to island environments and cultures

1.1 STEREOTYPES AND REALITIES

Our goal in writing this book is to examine relationships between island environments and people. We explore how these dynamic relationships have changed both the natural history of islands and the interactions between humans and nature. We approach this goal primarily from our perspective as ecologists who have lived and carried out research on islands. Humans both exploit and conserve natural resources, and human culture influences the attitudes of humans toward their environment. Recent surges in human populations on islands and the loss of the historical isolation of islands have removed many traditional restraints to over-exploitation of island resources. As a result, island ecosystems are extremely vulnerable to further damage. If the struggle to reconcile human use of resources and the maintenance of natural systems can be managed on islands, within their limited areas, then models developed here might guide those attempting to restore mainland systems.

Our image of islands is part stereotype, part reality. One popular stereotype is that of an ideal vacation spot. Often included in this idyllic vision are balmy temperatures, gorgeous sandy beaches, crystal-clear, turquoise water, and palm trees fluttering in the warm breezes. Colorfully clad tourists sit in wicker chairs in open-air bars and are served piña coladas by laid-back locals, while sensual music and the muffled sounds of surf fill the air. The implication of this stereotype might be that there is a freedom of knowing that we can visit, enjoy the experience, be relieved of all worries, and then return home. A second stereotype is a place of mysterious, often illegal action. This stereotype is filled with thoughts of pirates, drug runners, and action heroes chasing villains on narrow roads through coconut groves or across tranquil bays to secret hideouts. Inevitably, the locals speak some kind of pidgin and cavort with both the villains and their pursuers, but all action stops for the evening fish fry, cocktails, and languid dancing at sunset. The implication is that any shenanigans are local, perhaps temporarily exciting, but ultimately unimportant to visitors and therefore ephemeral. A third stereotype conjures islanders as out-of-touch, little-interested in modern intellectual or cultural pursuits, and still vaguely associated with a history of cannibalism and tribal warfare. In this view, island cultures are as isolated from modern culture as islands are physically isolated from the mainland. At worst, they are perceived as cultural backwaters or oddities, with few experiences worth sharing or valuing other than as curiosities. Generally implicit in each of these stereotypes is a harmony between humans and a benign natural world. There is plenty of local food, fresh water, entertainment, and space. Island realities share some features with these three stereotypes, at least on those islands that are tourist destinations. Certainly, many islands, including non-tropical ones, earn their reputations as lovely blends of quiet, natural beauty, friendly inhabitants, positive attitudes, and exotic cultural experiences (Fig. 1.1).

A different kind of stereotype is of culturally isolated islanders who develop a high degree of self-assurance, and believe that their society is more civilized than that of mainlanders. This attitude is encapsulated by a headline in an English newspaper declaring "Fog in

Fig. 1.1   Pier at Cromer, a seaside holiday destination in Norfolk, England.

Channel – Continent Isolated" or the term that Hawaiians have for the city of Las Vegas, Nevada (USA): "our ninth island." The island-based cultures of the British Isles and Japan transformed such self-assurance into political and ecological influences on a global scale. Implicit in this stereotype of islands and islanders is a lack of harmony between humans and nature. Nature is seen as an impediment that, like an enemy, needs to be dominated or destroyed before progress can be made. In this view, the natural world is a distraction or obstacle to be overcome. This perspective has been demonstrated by those who see islands as convenient places to mine, conduct weapons testing, construct military bases, build real estate empires, or otherwise invest in the artificial features of island communities (e.g., Hong Kong, Singapore).

Regardless of the stereotype, problems abound for islands and islanders. Human inhabitants of islands have always dealt with natural disturbances such as volcanoes, earthquakes, cyclones, tsunamis, and landslides; however, modern civilizations are exacerbating some of these disturbances and creating new ones. Agricultural runoff, over-fishing, and silt erosion from the land after deforestation degrade near-shore ecosystems. Intentional or inadvertent introduction of alien species (see Glossary for definition of terms) have severely depleted populations of many native island plants and animals. The natural world that used to supply water and food in abundance is being polluted and island aquifers are being overdrawn. Rising sea levels resulting from climate change have already rendered some low-lying atolls uninhabitable. Paradise is an illusion; the reality of islands less enticing. The response of many islanders has been to emigrate. How those islanders who remain address present-day realities and how they choose to deal with increasingly urgent environmental problems will be illustrative for all of humanity as it strives to live more sustainably.

1.2 SPECIAL FEATURES OF ISLANDS

Islands are by definition geographically isolated from other land masses and their limited geographical area is clearly defined. Many ecological and cultural characteristics follow as a consequence. Populations of plants and animals that do colonize islands eventually become limited by finite resources. They also evolve in isolation from larger gene pools and sometimes become richly divergent from the original invaders. The Galápagos Islands in the eastern Pacific Ocean are a famous example of how a wondrous variety of organisms has evolved in isolation from mainland habitats. The combination of isolation and geographical finiteness of islands provides an excellent laboratory in which to study links between humans and the natural world. Isolation and finiteness, however, leave island ecosystems extremely vulnerable to outside influences. In this chapter, we introduce each of these themes (isolation, finiteness, vulnerability, and evolutionary experiments) and examine them in more depth in subsequent chapters.

1.2.1 Isolation

The degree of isolation has a direct bearing on which species of plants and animals can colonize them and on the subsequent evolution of those colonizers. Continental islands were once a part of a mainland and are less isolated than oceanic islands, which have never been in contact with a continent. Islands are also not permanent geological features, but instead emerge and submerge and are shaped and reshaped by tectonic activity. Most terrestrial organisms rarely cross large bodies of water, so colonizing newly formed oceanic islands can take decades to millennia, depending on the degree of isolation. Typically, continental islands are more easily colonized. Remote islands can be colonized by birds and insects that can fly thousands of kilometers. Certain spiders and spores can be blown equally long distances. Some islands are colonized by organisms arriving on rafted plant matter or by terrestrial animals that swim. For marine organisms such as plankton, corals, algae, fish, eels, seals, or whales that utilize the marine habitats around island shores, colonization may be less problematic than for terrestrial organisms. Nonetheless, not all of these aquatic organisms are long-distance dispersers, so newly formed islands are only slowly colonized. For successful colonists that establish healthy populations on a previously uncolonized island, a rapid population expansion can occur, followed by evolutionary divergence into many new species. The future lineage, for those individuals that arrive without potential mates or that do not adapt easily to the island habitat, is more limited, with local extinction a common fate. Furthermore, those successful colonists now face the myriad challenges of subsequent human impacts (see Section 1.3).

1.2.2 Finiteness

Island size is finite and well-defined at large spatial scales by the shoreline and surrounding ocean. At smaller spatial scales, this boundary

Fig. 1.2   Cliff meets ocean at Waipi'o Bay on the Island of Hawai'i. Waipi'o Valley was the largest and most fertile valley on the Island of Hawai'i, once supporting thousands of people. A tsunami in 1946 and a flood in 1979 discouraged residents and currently only about 50 people live there.

between terrestrial and aquatic habitats can be sharp, as when cliffs meet the ocean (Fig. 1.2), but the boundary is more often fuzzy because habitats such as tidal pools, mangrove swamps, salt water marshes, and coral reefs complicate definitions of terrestrial and marine habitats. Such complex shorelines are often areas of rich biodiversity because of the many habitats they provide. However, the terrestrial and coastal marine habitats are limited. Therefore, populations of organisms, including humans, have clear limits to their growth.

When populations surpass their natural carrying capacity they can crash. A classic example of this occurred on St. Matthew Island in the Bering Sea. In 1944, the United States Coast Guard introduced 29 reindeer to provide an emergency food source for stranded sailors. With no predators and ample food, the reindeer population grew rapidly to about 6000 individuals until resources became limiting, the reindeer starved, and the population crashed (Fig. 1.3). Only 42 reindeer were alive in 1966. Survival for animals depends on keeping total resource consumption and population levels at or below the carrying capacity of the system. When that carrying capacity is exceeded, animals typically starve to death or die from other causes until the

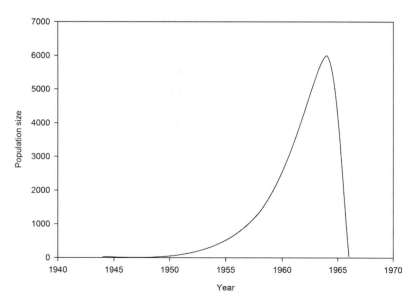

Fig. 1.3   Population size of the reindeer herd on St. Matthew Island, Alaska. See text for details on how this introduced population grew and then crashed when it ran out of resources. Redrawn from Klein (1966).

number of animals remaining utilizes the resources more sustainably. Under such conditions, local consumption must be reduced, either by intentional, organized means (including hunting, water and food rationing, sterilization, and more emigration than immigration) or by uncontrolled and chaotic events such as famine. Human populations are subject to the same limits to growth. Many islands now support human populations well beyond the carrying capacity of the island ecosystem because of a heavy dependence on imported resources.

1.2.3 Vulnerability

Island communities are particularly vulnerable to natural disturbances because of their finite physical extent and lack of escape routes for inhabitants. Volcanic islands can erupt, destroying all life, as happened on Krakatau, Indonesia in 1883. Earthquakes and huge, earthquake-induced landslides can cause parts of islands to collapse into the ocean. Cylones can damage forests and tsunamis can inundate coastal communities. When total destruction occurs, colonization depends on long-distance dispersal of organisms. Total destruction, however, is rare and most islands have remnant populations of plants and animals that can expand into the damaged areas. Large-scale, natural disturbances are the normal challenges that island organisms face.

People cause disturbances that often resemble natural disturbances (such as pavement surfaces that resemble lava) or trigger disturbances identical to those caused by natural phenomena (both road cuts and rain induce landslide formation). Many anthropogenic disturbances (those caused by people) augment the destructive power of natural disturbances. For example, it is likely that human effects on global climate have increased shoreline erosion due to sea level rise and increased cyclone intensity due to warmer oceans. However, some disturbances (including mine wastes, dynamiting of reefs by fishermen, or nuclear explosions) are so novel or are so toxic that organisms have no history of evolving in response to them.

Organisms evolve defenses only to survival problems. Unnecessary defensive traits are often lost (or never acquired) during long periods of evolution on isolated island ecosystems. Flightless, ground-dwelling birds have evolved on many islands (such as Tonga, Hawai'i, and New Zealand) where there were few predators. Island plants often lack protective features, such as spines or thorns, which are an adaptation to herbivores not found on islands. When egg predators, bird hunters, or herbivores eventually arrive, native species often cannot cope and become extinct. Humans can be the direct predators (as were hunters of moa in New Zealand). Alternatively, the predators can be introduced inadvertently (the Norway rat came to Hawai'i on European ships) or with a specific purpose, such as for biological control (stoats to control rabbits in New Zealand). Both inadvertent and purposeful invasions have occurred and continue to occur on many islands around the world. Similarly, human cultures on islands are also vulnerable to the invasion of new influences initiated by contact with the outside world.

1.2.4 Evolutionary experiments

Biological evolution

Islands can be seen as experiments in biological evolution and the closely allied process of cultural development. When a newly exposed island surface is initially colonized, biological diversity is low because those species that colonize an island are only a subset of the potential colonists from the nearby mainland. Over time, new species evolve on islands, eventually developing plants and animals unique to that island or island group. Organisms on neighboring islands or island groups can develop independently because isolation keeps genetic mixing to a minimum and the natural selection pressures that drive species evolution can differ between islands. The total diversity is limited by the finiteness of islands but promoted by their isolation and diverse habitats.

Habitat diversity on islands can be astounding and can change over time through such mechanisms as the formation, weathering, and the erosion of soils. Rugged topography (especially on volcanic islands) can also provide many niches for evolutionary divergence. By some accounts, the islands of Hawai'i contain 23 of the Earth's 24 terrestrial biomes. The prevailing trade winds on most mountainous tropical islands bring high rainfall to the windward sides while drying out the leeward sides. Varied climates promote diverse plant life that in turn favors many evolutionary adaptations by animals.

Cultural evolution

Human cultures on isolated islands have a parallel development of a rich variety of unique behaviors and attitudes. Sometimes these cultures provide insights into unusual attitudes toward the natural world, perhaps derived from the finiteness of resources and space or from diverse climatic conditions. For example, Hawaiian tribes were allotted slices of land from the shoreline up into the mountains, a clear recognition that each tribe needed resources that came from the entire elevational gradient. In another example, the crops that Māori settlers in New Zealand were able to grow depended on the latitude at which they settled. Cultural diversity is therefore as closely related to the natural history of islands as is biological diversity.

1.3 HUMAN IMPACTS

Human impacts on island ecosystems derive from our capacity to exploit many environments, our rapid population growth, and our energy- and resource-intensive lifestyles. Humans are highly capable of destroying natural habitats and habitat destruction is the most common cause of species extinction. We destroy habitats intentionally when we grow crops (Fig. 1.4), log forests, or fill in marshes to build resorts and parking lots. We also destroy habitats inadvertently

Fig. 1.4   (A) Sugar cane fields and the north flanks of the limestone ranges of the Cockpit Country, Trelawny, Jamaica; (B) derelict sugar mill on the Rio Cobre, St Catherine, Jamaica. Sugar cane has been a widespread agricultural introduction to tropical islands but has often followed a boom and bust cycle of production.

when we trigger forest fires, introduce invasive species that alter eco-systems, or cause the sedimentation of streams and reefs from the erosion of cropland soils. Urbanization and its associated activities (transportation and transmission corridors) now cover more than 5% of the Earth's surface and some islands, such as Puerto Rico, are highly urbanized. Increased wealth for islanders, whether derived from nat-ural resource extraction, tourism revenue, or remittances sent home from family members living abroad, increases the ecological footprint of communities that adopt a higher standard of living. The associated increases in consumption of imported luxury goods and of electricity, water, and space (for roads, houses, schools, or hotels) generally offset all but the most comprehensive of conservation initiatives. In some cases, however, wealth can reduce local ecological footprints, such as in Puerto Rico, where urbanization since the 1950s has resulted in abandonment of farmland and regrowth of forests. Human impacts on islands reflect human impacts on mainlands; but islands, unlike mainlands, do not have the physical space or resource base needed to buffer the rapid environmental changes that humans introduce. Efforts to avoid or mitigate human impacts on islands include conser-vation, management of invasive species, and restoration. These efforts are both aided and constrained by the finite space and resources avail-able on islands. On the one hand, removal of invasive species from an entire island and restrictions on entry of new ones is easier to achieve on small island land masses than on continents. On the other hand, islands have limited populations of rare organisms and a lack of alter-native habitats for organisms to move to when disturbed.

1.3.1 Conservation

Early human colonists of islands have a poor record of conserving native plants and animals. For example, Neolithic agriculture, wher-ever it was practiced, destroyed many local resources, often within only decades of its introduction. Even today, most islanders derive their food either from intensive, industrial-scale agriculture and aqua-culture or from imported food products, so their link to natural eco-systems is tenuous at best. There are many reasons to conserve species and habitats, which could include a developing sense of direct depend-ence on the land, but often involve broader-scale issues such as glo-bal rarity of a species, clean air, or clean water. The physical aspects of islands focus conservation issues in many ways. On the positive side, there are unique plants and animals that may garner worldwide attention and support; and, if conservation initiatives are funded properly, these organisms have a high chance of surviving. However, there are both biotic and political challenges to island conservation. Because there are no alternative habitats to rely on and there is little or no migration of new individuals, genetic and population-level problems of small populations must be addressed. In addition, there is often extreme pressure from many potential users for the limited areas that are still more or less in a natural state. Finally, despite the ample publicity about endangered island species and ecosystems, mainland conservation projects tend to be much better funded. For example, 31 of the 95 bird species listed under the United States Endangered Species Act live only in Hawai'i, and there are less than 1000 individuals remaining of 17 Hawaiian bird species. Nonetheless, only 4% of the total expenditure on restoring endangered bird species in the United States is directed toward the Hawaiian birds.

1.3.2 Invasive species and their management

Invasive plants, animals, and microbes are species that arrive, survive, thrive, and spread into a new habitat (Fig. 1.5). They are called "aliens" by those emphasizing their negative impacts, "exotics" by those who enjoy their novelty, and "non-natives" by those inclined to be neutral about their presence or who want more information about those impacts before deciding. All island organisms were originally invasive and no one has a clear definition of how long a species has to inhabit a given island in order to be considered a native. Curious parallels exist within modern human cultures. Thirty or more years of residency in New Zealand, Iceland, or Japan will not make you a native. One commonly used criterion is that native species are those present before human arrival. Another criterion is that non-native species are ones that arrived with human assistance, whether intentional or not.

Decisions about the desirability of an organism are mostly subjective. We do not tend to view humans as an invasive species, except in the rare cases where human access is restricted, such as on some island nature preserves. Characteristics of successful invaders tend to focus on how they resemble weeds (with high reproductive rates, short life spans, and wide ecological tolerances). However, invasive species research is in its infancy and we have only begun to study invasive species that are slower to colonize and have longer life spans. Some invasive species out-compete natives by usurping resources and inhabiting

Fig. 1.5   Introduced brushtail possum in a whārangi tree, Puketī Forest, North Island, New Zealand. Brushtail possums are native to Australia and were introduced to New Zealand by Europeans to establish a fur industry but have caused much damage to New Zealand's plants and animals.

new niches. Others succeed by altering ecosystem processes such as nutrient cycling or water balance. Often, invasive species promote the spread of other invasive species such as the Japanese white-eye bird that commonly disperses invasive trees, such as fire tree and straw-berry guava, on the Island of Hawai'i. However, these white-eye birds also disperse seeds of native plants species where native plants still exist.

Humans manage invasive organisms in several ways. Preventing invasions is possible with strict quarantines, but these can be breached, even on islands where most access is through airports or boat docks. Eradicating invaders soon after colonization is ideal but often unsuc-cessful because many invaders are cryptic or rare during the earliest stages of invasion; thus, by the time eradication is attempted, invasive populations are often well established. Removal of plants (pulling, spraying) or animals (trapping, shooting, poisoning) can help stem the tide but complete success means addressing and blocking the corridor for invasion. Complete removal of invasive animals is tractable on some islands and once eradicated, future invasions can be more easily controlled than on mainlands. In New Zealand, it has been observed that "conservation is all about killing." In one New Zealand example, on Kapiti Island, intensive trapping and poisoning removed all invasive rats, cats, and brushtail possums; native birds are currently thriving. Human visitors are carefully screened so that they do not inadvertently introduce more non-native mammals. Other removals have been more problematic. Removing the big-headed ants that plagued native seabirds from several small islands off Oʻahu, Hawaiʻi resulted simply in the invasion of the even more noxious crazy ant. Nonetheless, many island systems offer excellent case studies that can then guide the more difficult challenge of removing invasive species from mainland ecosystems.

1.3.3 Restoration

The recovery of damaged habitats and diminished populations is a big challenge that requires creative minds, ample resources, and a clear set of goals. Little is known about how to define or re-create healthy ecosystems, so most restoration goals are very limited. A typical approach is to focus on re-establishing certain dominant native species and then hope that other species and associated ecosystem functions will return unaided. Restoration on islands is potentially easier than on the mainland for several reasons. The initial cause of the damage may be more easily isolated and addressed. In addition, islands usually offer less interference from migrant animals and weedy species than the mainland. On small islands, there is the exciting possibility of restoring whole ecosystems. Finally, the intimate link many islanders have with their natural environment and islanders' subsequent recognition of the immediate benefits of restoration can provide a supportive environment for restoration projects. However, there can be divergent goals for restoration projects spanning the modern social structure of many islands. Goals for some groups, especially indigenous people on islands, may be about restoring plant or animal populations for sustainable use by people, whereas other groups have preservation of species or ecosystems as goals.

1.4 OUR ISLAND FRAMEWORK

We explore the themes of isolation, finiteness, vulnerability, and evolutionary experiments by focusing on nine island groups or clusters where we have worked, six of which we have lived on. Other islands will be mentioned but our story stems from our personal experiences with the biological and cultural challenges of these nine island groups. These include (in increasing order by size) Tonga, Canary Islands, Puerto Rico, Jamaica, Hawai'i, Iceland, New Zealand, the British Isles, and Japan (Table 1.1 and Fig. 1.6). In addition to one or several large islands, each group includes several to thousands of smaller islands (see Chapter 2). These island groups cover a wide range of latitude (52 degrees south to 67 degrees north) and longitude (180 degrees east to 177 degrees west) and are found in the Atlantic and Pacific Oceans and the Caribbean Sea. Some were settled tens of thousands of years ago (British Isles) while others were discovered and settled by humans within the last 1000 years (New Zealand). Current human populations range from 112 000 (Tonga) to 127 433 000 (Japan) with population densities ranging from 3 (Iceland) to 439 (Puerto Rico) people per km^2. Polynesian, Asian, European, and African cultures are represented. Iceland, Hawai'i, and New Zealand still have large indigenous cultures developed by people who arrived about 1160 years (Iceland), 1200 years (Hawai'i), and 730 years (New Zealand) ago. While Icelanders are still very homogeneous in their genetic and cultural background (European), the original Polynesian residents of Hawai'i and New Zealand have been augmented by mixtures of Polynesian, Asian, and European cultures.

This set of island groups provides a diverse template for us to explore our four themes of isolation, finiteness, vulnerability, and evolution and to describe how each group is unique or similar to the other groups. With a wide representation of oceans, climates (Figs. 1.7 and 1.8), geological origins, geographical shapes and sizes, biomes, organisms, and human population densities (Figs. 1.9 and 1.10), we can look for generalizations about island phenomena. Our focus in this book will be mostly on terrestrial ecosystems. However, a discussion of island ecosystems cannot disregard interactions with near-shore aquatic habitats that are intimately linked to terrestrial processes, such as reef damage from cyclones or soil erosion, and fishing activities of islanders.

Table 1.1. *Geographic and demographic information about the nine island groups that form the focus for this book (listed by increasing area)*

Island group	Area (km²)	Latitudinal range (degrees)	Longitudinal range (degrees)	Population (1000s)	Population density (number per km²)	Human settlement (approximate date)
Tonga	748	15–22 S	173–177 W	112	162	800 BC
Canary Islands	7 447	27–29 N	13–18 W	1 996	268	500 BC
Puerto Rico	9 104	18 N	65–68 W	3 994	439	1 800 BC
Jamaica	10 991	18 N	76–78 W	2 804	255	AD 700
Hawai'i[a]	16 421	16–23 N	154–162 W	1 288	78	AD 800
Iceland	103 125	63–67 N	13–24 W	306	3	AD 874
New Zealand[b]	268 680	29–52 S	166–180 E	4 292	16	AD 1280
British Isles[c]	315 134	49–61 N	8 W–2 E	64 926	206	780 000 BC
Japan[d]	377 873	24–45 N	123–144 E	127 433	337	130 000 BC

[a] Excludes Emperor Chain.
[b] Includes Chatham Islands, Kermadec Islands, and sub-Antarctic islands.
[c] Includes the UK, Ireland, Shetland, Hebrides, and Isle of Man but not Channel Islands.
[d] Excludes the Kuril Islands.

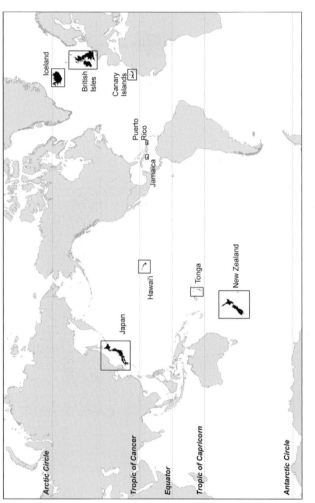

Fig. 1.6 The location of nine island groups where the authors have worked. These island groups form the background for the ideas presented in this book.

Fig. 1.7   Iceland has a cool-temperate climate with many glaciers, such as at Jökulsárlón along the south coast.

Fig. 1.8   A rainforest in the tropical Luquillo Experimental Forest, Puerto Rico. This intact forest is the only tropical National Forest in the United States and is a popular weekend destination for residents of urban San Juan.

1.5   SCOPE

Our first two themes (isolation and finiteness) are addressed in Chapter 2 in which we describe the geology, geography, climate, and soil conditions of each of the nine island groups. The discussion of geology is organized by island group while the other three topics are

Fig. 1.9   Rural housing and deforested slopes in the Hope River Valley in northeast Kingston, Jamaica.

Fig. 1.10   The densely populated island of Oʻahu, Hawaiʻi includes the city of Honolulu.

addressed thematically, with island groups intermingled. We discuss five types of islands (oceanic, atoll, continental, continental fragment, and anthropogenic).

How natural disturbances impact islands and how native plants and animals react to disturbance are the topics of Chapter 3, which addresses our third theme (vulnerability) in areas minimally influenced by humans. Islands are affected by a fascinating variety of disturbances, some resembling mainland coastal disturbances (such as cyclones, coastal erosion) and some that are more commonly associated with islands (such as the effects of bird colonies). Our sample of island groups represents an array of disturbance regimes that provides the structure for our thematic approach to the topics in this chapter.

We explore the first aspect of our fourth theme (islands as evolutionary experiments) in Chapter 4 with a discussion of the unique plants and animals of islands. We discuss how they disperse, radiate into many new species, evolve a wide variety of sizes, and lose traits, such as the ability to fly. Some of these traits make them highly vulnerable to extinction. Again, we use examples primarily from the nine island groups but approach the topic by theme rather than by island group.

The second half of theme four (islands as cultural experiments) is introduced in Chapter 5 (organized by island group) where we discuss the initial human dispersal to the island groups and how these colonization events progressed. In Chapter 6 (organized by island group) we integrate all four themes with human activities for the period following initial human colonization until modern times. Chapter 7 (organized by theme) covers modern human impacts (1950–2000) that coincide largely with the end of isolation following the introduction of widespread air transport.

In Chapter 8, we discuss the future of islands in the context of two of their biggest challenges: overpopulation and climate change. We offer some practical solutions to these challenges because how humans respond to them will determine the future of island ecosystems and cultures.

SELECTED READING

Crosby, A.W. (2004). *Ecological Imperialism: The Biological Expansion of Europe, 900–1900,* 2nd edn. Cambridge: Cambridge University Press.
del Moral, R. and L.R. Walker (2007). *Environmental Disasters, Natural Recovery and Human Responses.* Cambridge: Cambridge University Press.

Diamond, J. (1999). *Guns, Germs and Steel: The Fate of Human Societies*. New York: Norton.

Dunn, P.D. (1999). *Environmental Change in the Pacific Basin: Chronologies, Causes, Consequences*. New York: Wiley.

Flannery, T. (1994). *The Future Eaters*. New York: Grove Press.

Flenley, J. and P.G. Bahn (2003). *The Enigmas of Easter Island: Island on the Edge*, 2nd edn. Oxford: Oxford University Press.

Klein, D.R. (1966). The introduction, increase, and crash of reindeer on St. Matthew Island. http://dieoff.org/page80.htm (accessed 22 October 2010).

Leonard, D.L., Jr. (2008). Recovery expenditures for birds listed under the US Endangered Species Act: The disparity between mainland and Hawaiian taxa. *Biological Conservation* **141**: 2054–61.

Rolett, B.V. (2007). Avoiding collapse: pre-European sustainability on Pacific islands. *Quaternary International* **184**: 4–10.

Yoffee, N. (ed.) (2009). *Questioning Collapse*. Cambridge: Cambridge University Press.

2

The physical setting

2.1 ISOLATION AND FINITENESS

The mystique of islands is in part due to two physical features: their isolation and their sharply defined boundaries or finiteness (Fig. 2.1). In this chapter, we discuss the physical setting of the nine island groups to provide the background needed to interpret the natural disturbances to which they are subjected (see Chapter 3) and distributions of their plants and animals (see Chapter 4). We discuss the geological origins of each island group and cover the geographic relationships among islands in each group, their topographical relief, and the nature of the coastline; each of these parameters is of key relevance to colonizing organisms, from small barnacles to humans looking for harbors. Next, we discuss the climate of the island groups, including the reasons for variation of precipitation and temperature within islands. We finish by describing how soils are shaped by geological, topographical, climatic, and biological influences on the islands.

2.2 GEOLOGY

Oceanic islands have never been in contact with a continent and are largely created when volcanoes form from magma extruded through hotspots or weak zones in crustal plates (Plate 1). Uplifted coral and limestone formed on a volcanic base can also form oceanic islands. Oceanic islands of volcanic origin form when mountains emerge above the sea after extensive deposits of lava. Subsequent movement of the plates away from the hotspots allows the formation of new islands and over time this movement results in island groups. Continental islands are an emergent part of a continental shelf with relatively

21

Fig. 2.1   Sea cliffs at Yesnaby, Mainland, Orkney, Scotland.

shallow water between them and the continent. Land bridges often connect continental islands to the continent during periods of low ocean levels, such as during glacial advances. Continental fragments are islands that separated millions of years ago by plate movements from a continent and are now surrounded by deep ocean water. Four of the nine island groups we consider in this book are oceanic, one is part of a continental shelf, and four are continental fragments (Table 2.1). These nine island groups represent a wide range of ages (new volcanic surfaces to rocks more than two billion years old). Although the continental islands are generally older than the oceanic islands, the range of ages of islands within an island group can also vary widely (Table 2.1). The current shape of islands is only a snapshot of the dynamic processes of land formation, subsidence, uplift, and sea-level changes.

2.2.1 Oceanic islands

About one million submarine volcanoes exist but only several thousand have reached the surface of the ocean to form oceanic islands. Oceanic islands may form within crustal plates (large pieces of the Earth's crust that move independently) as clustered (Canary Islands) or linear groups (Hawai'i), or along plate boundaries where one plate pushes under another in a process called subduction (Tonga and Iceland). Oceanic islands tend to be shorter-lived than continental islands because they subside back into the ocean, although this

Table 2.1. *Physical characteristics of nine island groups (see Table 1.1 for further descriptions). Coastline length used a 1 km scale. My = million years*

Island group	Ocean or sea	Geologic origin	Age (My)	Isolation (km to nearest mainland country)	Maximum altitude (m)	Coastline (km)	Coast/area ratio (km/km^2)	Precipitation range (mean annual; mm)	Temperature range (mean annual; °C)
Tonga	South Pacific	Oceanic	0–6	3460 (Australia)	1 033	419	0.560	1 610–2 930	23–28
Canary Islands	North Atlantic	Oceanic	0.05–20	96 (Western Sahara)	3 718	1 500	0.201	112–594	16–24
Puerto Rico	Caribbean	Continental fragment	17–190	1 056 (Venezuela)	1 328	700	0.055	819–1 755	20–27
Jamaica	Caribbean	Continental fragment	17–190	663 (Honduras)	2 256	1 022	0.093	800–1 505	21–27
Hawai'i	North Pacific	Oceanic	0.4–74	3 824 (USA)	4 205	1 200	0.073	168–6 147	5–24
Iceland	North Atlantic	Oceanic	0–15	1 328 (Norway)	2 110	4 988	0.048	321–2 716	0–5
New Zealand	South Pacific	Continental fragment	70	2 153 (Australia)	3 754	15 134	0.056	495–6 337	8–15
British Isles	North Atlantic	Continental shelf	55–2 700	34 (France)	1 343	18 429	0.058	551–2 008	8–12
Japan	North Pacific	Continental fragment	15	200 (Korea)	3 776	29 979	0.079	800–7 000	6–21

Fig. 2.2   Uplifted limestone beds along the west edge of Kenutu,
a small island on the eastern side of the Vava'u group, Tonga.

process can be delayed in tropical regions by the formation of atolls
(rings of coral) around the island.

Tonga

The islands of Tonga are a mixture of three types of formation: islands
that are purely volcanic; islands composed of limestone on a volcanic
base (Fig. 2.2); and islands formed from uplifted coral. The volcanic origin of the islands comes from subduction of the Pacific Plate under the
Indo-Australian Plate during the last six million years. This subduction
occurs along the more than 10 000 m deep Tonga Trench, the second
deepest place in the world's oceans (after the Marianas Trench south
of Japan). The islands of Tonga occur along two parallel, north–south
lines on the Indo-Australian plate, just west of the trench. Currently,
this subduction is occurring at a rate of 1–1.7 cm per year and there
are frequent eruptions as a result, particularly in western Tonga.
In this region, there are about 36 sunken volcanoes and emergent
islands (such as Kao, Tofua) as well as several new islands (Late'iki,
which appeared in 2006, about 20 km southwest of Late Island, and
Fonuafo'ou). In 2009, a new underwater volcano erupted 10 km southwest of Tongatapu. This new volcano sent steam, ash, and smoke
more than 100 m into the air but did not produce a permanent island,
unlike another 2009 eruption 60 km northwest of the Tongan capital
of Nuku'alofa that did produce a permanent island.

The islands of Tonga not directly created by new volcanism are caused by crustal uplift as they subduct toward the west owing to the mass of the new volcanic crust. These islands (including 'Eua, Tongatapu, and the Vava'u Group) have steep cliffs on their eastern and northern shores while their western shores are submerging, thereby creating many islets and lagoons. A third group of islands (Ha'apai Group) consists of eroding coral atolls and a raised barrier reef.

Canary Islands

The Canary Islands are also volcanic in origin and have never been connected to mainland Africa. Their formation is due to a combination of a hotspot (a zone of upwelling of hot magma) and the collision of crustal plates. The seven main islands that exist today were formed between 0.05–20 million years ago but in no particular spatial order. Volcanic activity last occurred on the oldest island of Fuerteventura about 4000 years ago, but 5 of the 12 (Lanzarote, La Palma, El Hierro, Gran Canaria, and Tenerife) have been volcanically active in the last 3000 years. The most recent eruption was in 1971 from Cumbra Vieja on La Palma. During the geological history of the Canary Islands, the exposed surface area of the islands has varied twofold depending on catastrophic landslides, subsidence, and sea-level changes.

Hawai'i

The islands of Hawai'i consist of two arms of a chain of islands that have formed over a hotspot. Volcanoes form as the Pacific Plate moves across the hotspot at a rate of about 9 cm per year. The southeastern and youngest arm of the island chain begins with the still-submerged Loihi volcano that is 30 km south of the Island of Hawai'i and within 1000 m of the ocean surface. Eight islands in the Hawaiian chain now rise more than 400 m above sea level, with successively older islands to the northwest. The youngest is the Island of Hawai'i, which has several active volcanoes (Fig. 2.3). Mauna Kea Volcano, which rises 10 200 m from the bottom of the ocean, is the tallest "mountain" in the world. On its slopes, Kīlauea Volcano is still growing from lava flows spilling into the surrounding ocean (Box 2.1). The oldest flows on the Island of Hawai'i are 0.5 million years old. Maui, Kaho'olawe, Lāna'i, Moloka'i, and O'ahu were likely at one time a single giant island that formed 1.2–3.0 million years ago. Kaua'i and Ni'ihau were also connected and formed 4.7–5.1 million years ago. Further along the chain, Kure was

Fig. 2.3   Recent eruption (2009) of Halema'uma'u Crater of Kīlauea
Volcano on the Island of Hawai'i. Steam and sulphuric acid combine to
form a sometimes deadly vog or volcanic fog. Note the broad, nearly
flat contours of this young shield volcano.

formed 29.8 million years ago and the final island of the initial chain
is Daikakuji Seamount, formed 43 million years ago. It is 3493 km
northwest of the Island of Hawai'i. The Emperor Chain (currently submerged) extends northward another 2327 km, culminating in Suiko
Seamount (70–80 million years ago) and Meiji Seamount, and is the
longest chain of seamounts in the world. The different orientation
of this chain of ancient islands suggests that the direction of crustal
movements has changed in the past.

Box 2.1. Flow, lava, flow

John Kjaergaard needed an assistant and I needed adventure, so
the two of us traveled to active lava flows on the Island of Hawai'i
between 1985 and 1987. Several pairs of boots were ruined as we
crossed hot and sometimes glowing lava so that John could film
flaming forests, rivers of lava, lava tubes, ash plumes, crater lakes
(where the lava moved like crustal plate tectonics), and fountains
of lava flowing into the ocean. Since 1983, eruptions from the
most active vents from Kīlauea Volcano have added 2 km² of land
to the island, produced a 250 m tall cinder cone, formed a new
lava lake, and covered 117 km² of forests. LRW

As the islands move off the hotspot they are eroded and short-ened, due to contraction as they cool. This contraction leads to subsid-ence or a lowering of the island's overall height above the ocean and it is initially as rapid as 2.5–3 mm per year on the Island of Hawai'i. The older the islands are, the more slowly they subside. Seismically trig-gered submarine landslides have eroded the flanks of the volcanoes and cover more area than five times the current terrestrial surface of all the islands combined.

Iceland

Iceland has formed from volcanic activity over a hotspot on the mid-Atlantic ridge, where the North American and Eurasian plates are spreading apart at a rate of about 3 cm per year. Iceland is estimated to be 15 million years old. The island is divided into northwest and southeast halves by the fissures and volcanoes. One famous 60 m wide fissure (Fig. 2.4) has provided a dramatic backdrop for the gathering place (Thingvellir) of the Icelandic government since AD 930. Geysers, hot springs, recent lava flows, and even new offshore islands (Box 2.2) attract geologists, biologists, and tourists to Iceland. There are cur-rently seven active volcanoes (two of which are beneath glaciers) and more than 700 hot springs.

Fig. 2.4   Central spreading rift of Iceland near Thingvellir. In the valley created by this rift, the first Icelandic government was formed and continued to meet there for hundreds of years.

> **Box 2.2.** A brand new island!
>
> In 1963, Surtsey became the youngest island in the world. After
> four years of erupting, it reached a total height of 174 m and
> a maximum area of 2.7 km². Today, following rapid erosion of
> columnar basalt cliffs on its southern shores, the island is only
> 1.4 km² and is no longer the youngest island. The island has a
> central core of ash that is more resistant to erosion than the
> basalt, so the island will probably be around for a long time to
> come. I visited Surtsey in 2003 with Icelandic biologists who are
> allowed on the island for only several days each year. The team
> tracks the invasion and fate of the 14 species of breeding birds,
> 69 species of vascular plants, and several hundred species of
> terrestrial invertebrates that have colonized since the eruption.
> Except for some geologists and a hut maintenance crew, no one
> else is allowed on this island that is now protected as a UNESCO
> Natural World Heritage Site. This protection gives scientists a
> rare opportunity to study a new land surface minimally altered
> by humans. LRW

2.2.2 Continental islands and land bridges

British Isles

The British Isles are our only example of an island group on a continental shelf. Their geological history is very complex and fascinating because it spans half the history of the Earth and involves much movement of various pieces of the British Isles around the globe. The oldest rocks are metamorphic gneisses at least 2700 million years old found in the extreme northwest of Scotland and the Hebrides. Younger rocks are generally found as one moves south and east in the British Isles. The next oldest rocks are sedimentary sandstones from about 1000 million years ago, followed by a mixture of sedimentary and volcanic rocks formed about 600 million years ago when England and Scotland were part of the large, super-continent Gondwana and were located 60 and 20 degrees south of the Equator, respectively. Several mountain-building periods and intervals of erosion and sandstone deposition followed, as pieces of today's British Isles rafted on crustal plates, experiencing submersion and re-emersion from under shallow seas, arid climates in the middle of continents, and continued cycles of mountain building (especially 600 and 280 million years ago) from plates colliding and subsequent erosion. The Atlantic Ocean and the North Sea formed and multiple inundations created shale, North Sea oil reserves, and chalk deposits. The last volcanic activity was about 55–60 million years ago. Ocean levels have fluctuated more than 100 m during the last 25 million years, generally rising during interglacial periods when the water in the polar ice caps melted. The British Isles have often been a part of the same continental plate as mainland Europe but land connections have come and gone. Uplift ended about two million years ago with generally modern landscape features in place. The current isolation of the British Isles occurred between 6000 and 10 000 years ago as ice sheets receded by half and the North Sea rose about 40 m, severing land bridges. Presently, the geological dynamism of the British Isles continues: Scotland has been rising since the removal of the weight of the ice sheets from the last glacial period, and London has been sinking from compaction of clay-rich soils underneath it.

2.2.3 Continental fragments

Puerto Rico and Jamaica

Puerto Rico and Jamaica share a remarkable geological history that is still poorly understood. They are composites of volcanically derived rocks, granite intrusions, and limestone from deposition of the calcium in ancient animal shells. Most evidence suggests that they are a part of a continuous arc of land, which emerged in the Pacific Ocean about 130 million years ago after the breakup of Pangaea into Laurasia and Gondwana. This arc then collided with the Bahamas Platform of the North American Plate, triggering a spreading of the North and South American Plates and the movement about 1200 km eastward of the arc to its present position in the central Caribbean Sea during the last 50 million years. In essence, these islands drifted through the Isthmus of Panama long before the Panama Canal was built!

During its eastward movement, this arc, called the Caribbean Plate, was mostly under water; individual islands of today were likely submerged during much of this migration (Jamaica was submerged for 20 million years). Jamaica and the southern part of Hispaniola followed the leading edge of the arc that is present day Cuba, Puerto Rico, and the northern part of Hispaniola, by at least 10 million years. Several pieces of the Caribbean Plate were therefore moving relatively independently; the Lesser Antilles were the last islands to reach their current locations. The Greater Antilles, including Puerto Rico and Jamaica, appear to have moved both northeast and west, rotating along the plate boundaries. Puerto Rico separated from Cuba about 39 million years ago and from northern Hispaniola about 5 million years later. Jamaica has most likely remained spatially independent from other Caribbean islands during its last 50 million years. The oldest rocks in Puerto Rico (about 130 million years old) therefore originated in the Pacific Ocean. Volcanic activity on both islands peaked between 70 and 17 million years ago, interspersed with periods of erosion, uplift, and formation of limestone hills from sediment accumulated while the islands were underwater. The current form of the islands was determined about 12 000 years ago. Today, Puerto Rico and Jamaica, along with Hispaniola, are still seismically active because they sit at the southwestern edge of the North Atlantic Plate that is subducting under the lighter Caribbean Plate along the >8000 m deep Puerto Rico Trench 115 km north of Puerto Rico.

New Zealand

Some of the rocks present in New Zealand today formed when it began to break away from Gondwana as a new continent called Zealandia about 85 million years ago. During the 30 million years New Zealand took to reach its present location 2000 km east of Australia, coal and limestone were formed when New Zealand was submerged. By 25 million years ago, the Pacific Plate started subducting under the Australian Plate (Fig. 2.5) and new mountain building began with uplift and volcanic activity between 5 and 18 million years ago. During the last 1.8 million years, the Southern Alps rose and now contain the highest peaks, which are >3000 m tall. Large volcanic eruptions (such as Taupo; see Chapter 3) and multiple glaciations carved many of the current contours of New Zealand. During glacial advances, sea level dropped as much as 100 m, resulting in expanded land surface and joining of the North and South Islands. New Zealand straddles the dynamic boundary of the Australian and Pacific plates, called the Alpine Fault. The resultant pressures at this fault are buckling New Zealand and causing some of the most rapid rates of uplift in the world (2.5 cm/year), which are accompanied by fast erosion rates and frequent earthquakes. Lateral movements along the Alpine Fault triggered by earthquakes can reach up to 8 m at a time. On-going volcanism is also a feature of the collision of the two plates. The largest city, Auckland, is nestled among volcanoes that last erupted from 50 000 to just 600 years ago. During a 7.8 moment magnitude earthquake

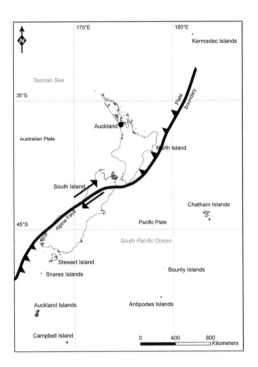

Fig. 2.5 Plate boundaries in New Zealand and the Alpine fault.

in 2009, the southwestern tip of New Zealand moved 30 cm closer to Australia. These events are enough to keep every New Zealander aware that few features of the landscape can be considered stable.

Japan

The islands of Japan were created about 15 million years ago when they separated from the eastern coast of Asia as the Sea of Japan formed from subsiding of the Earth's crust. The geological foundation of Japan is ancient gneiss and granite rocks that were formed 2.5–3.8 billion years ago. Tuffs, sandstones, and slates formed 250–540 million years ago and granite intrusions (Fig. 2.6) were common 65–250 million years ago. Volcanism dominated the last 65 million years of Japanese geological history. About 200 volcanoes have erupted in the last 2 million years in Japan, half of which are considered active (erupted within the last 10 000 years). Japan therefore has a remarkable 10% of the world's active volcanoes (Plate 2) because it lies at the junction of four tectonic plates (Fig. 2.7) where both subduction and uplift occur. The northern half of

Fig. 2.6   Mochumo-dake (994 m) and adjacent high granite peaks, near
Onoaida, Yakushima, Japan.

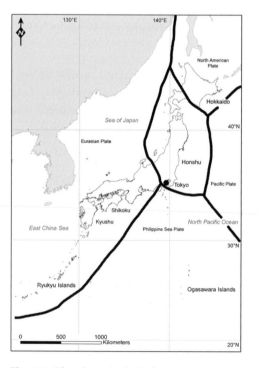

Fig. 2.7   Plate boundaries in Japan.

Japan is on the North American Plate and the southern half is mostly on the Eurasian Plate. The Pacific Plate subducts under the North American Plate along the Kuril and Japan Trenches and under the Philippine Sea Plate along the Izu-Bonin Trench. The Philippine Sea Plate, in turn, subducts under the Eurasian Plate along the Suruga-Nankai and Ryukyu Trenches. This last subduction zone is the most recent.

2.2.4 Barrier islands

Barrier islands occupy 12% of open ocean coastlines and there are several thousand of them worldwide. They are made of unconsolidated sand or gravel, and can move as storms deposit or remove sands, particularly during cyclones. Barrier islands are longer than they are wide, with open ocean on one side and a lagoon on the other. There are four main types. Coastal plain barrier islands lie along the flat coastal plains on the shores of the North Atlantic and Gulf of Mexico. Arctic barrier islands are shaped by waves, ice, and permafrost melt waters. Delta barrier islands line estuaries and deltas from rivers. Barrier islands associated with glacial outwash plains (sandur) are the only ones that migrate seaward and are most commonly found in Iceland. Of the nine island groups, Iceland has 22 barrier islands (total length 187 km or 0.90% of its shoreline); New Zealand has one (length 27 km, 0.12%); and the British Isles have four (three in England [14 km; 0.07%], and one in Scotland [10 km, 0.04%]).

2.2.5 Anthropogenic islands

Humans make islands for a variety of reasons, both intentional and not. Sometimes their ownership becomes quite contentious (Box 2.3). Historically, constructed islands include medieval crannógs built in Ireland and Scotland (mostly in lakes but also sometimes in estuaries) that were likely built to provide a safe place not easily accessible by one's enemies. The island of Dejima was built off the coast of the city of Nagasaki, Japan in 1634 to allow Portuguese and later the Dutch (1640s–1840s) to trade in silk, wool, glassware, deer pelts, shark skins, and other products with the Japanese without desecrating Japanese soil in the period when access by foreigners was strictly regulated. In the 1850s in Tokyo Bay, an island called Odaiba was built in an unsuccessful attempt to protect Tokyo from the United States Navy, under Commodore Perry, that was forcing Japan to trade with countries other than the Dutch. Today, Odaiba is a tourist zone with hotels, shopping malls, and convention centers.

> **Box 2.3.** Micronations
>
> Micronations are curious phenomena, where individuals or groups
> proclaim the establishment of a new country, issue currency
> and passports, and defend their territorial waters, but are not
> recognized by other countries. The best known micronation is
> perhaps Sealand, established in 1967 off the East Anglia coast of
> England on an abandoned gun platform and adjacent sand bar by
> a radio broadcaster and former English army major named Paddy
> Bates. He (and others claiming leadership of this micronation)
> have successfully fended off the British Navy and legal maneuvers
> to remove them as they are deemed to be in international waters.
> An even earlier micronation was created by Leicester Hemingway,
> the brother of Ernest Hemingway, in 1964 off the coast of Jamaica,
> using a barge and bamboo platform. This micronation was called
> New Atlantis and was meant to further marine research but it
> lasted only a few years before being destroyed by a cyclone and
> subsequent pillaging by fishermen. Perhaps the most quixotic
> example is of an artificially expanded reef near Tonga. After
> importing sand from Australia in 1971 in order to connect several
> small cays and build up the reef, a rich American from Nevada
> named Michael Oliver tried to establish a new, libertarian country
> called Minerva. It lasted only a few months before Tonga enforced
> its claim to the reef, but there have been several subsequent, less
> serious attempts to reestablish the micronation of Minerva; Fiji
> also claims the reef. Ironically, what seems to be missing from
> this political tug-of-war is discussion of the damage to the reef and
> fishing resources from the addition of the sand. LRW

Modern construction of artificial islands often occurs in order to
support airports. O'ahu in Hawai'i has a long runway built by filling in
a reef to accommodate jet travel. Japan has built five airports on artificial islands or extensions of natural islands, and more are planned.
The first was the Kansai Airport, a 10 km^2 artificial island built 3 km
offshore and completed in 1994. It has survived a major earthquake
and a typhoon but continues to subside into the silty ocean floor at
higher than expected rates. However, the sinking may now be stabilizing and the installation of adjustable girders to prevent further sinking will help address this problem. Construction lessons from Kansai
have been applied to other island airport projects in Japan and elsewhere. For example, the airport at Chubu near Nagoya in Japan has

curved shores to promote flow of original currents and naturally slop-ing edges to encourage use by wildlife. Another reason to build islands is to provide additional land for military bases. England has gradually expanded Whale Island in Portsmouth Harbor on the southern coast from a small, natural island to provide additional land for the Royal Navy. Finally, islands can be used in restoration activities. Puerto Rican engineers are building an artificial reef offshore from San Juan to replace coral reefs damaged by the grounding of a barge in 1994.

Unintentional islands are created as a byproduct of human activ-ities, including mine or urban waste deposition. Ironically named Dream Island (Yume no Shima) in Tokyo Bay, Japan, was created from urban refuse but is now covered with soil and has sports fields, a museum, a greenhouse, and a yacht harbor, as well as sanitation facilities.

2.3 GEOGRAPHY

The nine island groups represent the variation typical for both the isolation and the finiteness of islands (Table 2.1). The British Isles on a continental shelf are closest to a mainland (34 km from France) followed by an oceanic group (Canary Islands, 96 km from Western Sahara). The most distant from a mainland are two other oceanic groups (Hawai'i and Tonga; >3400 km from the USA and Australia, respectively). Iceland (an oceanic group) and the four groups that are continental fragments (Puerto Rico, Jamaica, New Zealand, and Japan) are all 200–2200 km from a mainland. Thus, with the exception of barrier islands that are always adjacent to large land masses, the geological origin of an island group is no indication of its degree of isolation.

Finiteness can be measured as the well-defined boundary of an island within the sea and this definition serves to separate mainland from island terrestrial habitats. However, the finite edge of an island also is blurred when one considers how the shapes of islands and their groups vary at both geological time scales as a result of plate tectonics and at glacial and interglacial time scales due to sea-level changes and isostatic rebound. Low sea levels can result in land bridges between islands and adjacent mainlands. The fluctuating shapes of islands have important implications for colonization of islands and subsequent evolution of plants and animals (see Chapter 4). In this section, we examine the spatial patterns, including coastlines and topography of the nine island groups, to understand better the physical template for the evolution of plant and animal communities.

Table 2.2. *Major islands within each of nine island groups. The total number of islands varies with minimal size criteria for an island (typically >0.1 km², sea level, and other factors so these numbers can be considered approximate. Inhabited refers to the existence of a permanent settlement*

Island group	Number of islands			Major islands (in descending order by area)	
	Total	>10 km²	Inhabited	>100 km²	>10 km²
Tonga	178	10	36	Tongatapu, Vava'u	'Eua, Niuafo'ou, Tofua, Late, Niuatoputapu, Foa, Kao, Lifuka
Canary Islands	12	9	8	Tenerife, Fuerteventura, Gran Canaria, Lanzarote, La Palma, La Gomera, El Hierro	La Graciosa, Alegranza
Puerto Rico	9	4	3	Puerto Rico	Vieques, Mona, Culebra
Jamaica	4	1	1	Jamaica	None
Hawai'i	132	8	7	Island of Hawai'i (Big Island), Maui, O'ahu, Kaua'i, Moloka'i, Lāna'i, Ni'ihau, Kaho'olawe	None
Iceland	17	2	4	Iceland	Heimaey
New Zealand	>100	29	16	South, North, Stewart, Chatham, Auckland, Great Barrier, Resolution, D'Urville, Campbell, Adams	19 additional islands
British Isles	>6 000	57	138	England + Scotland + Wales, Ireland, Lewis & Harris, Skye, Shetland, Mull, Anglesey, Islay, Man, Orkney, Arran, Wight, Jura, South Uist, North Uist, Yell, Achill, Hoy, Bute, Unst, Rum	35 additional islands
Japan	>6 000	36	426	Honshu, Hokkaido, Kyushu, Shikoku, Okinawa, Sado, Amami Oshima, Tsushima, Awaji, Yakushima, Tanegashima, Iriomote, Dogo, Rishiri, Miyako-jima, Shodoshima, Okushiri, Iki, Suo-Oshima	26 additional islands

Uplift due to various tectonic activities, generally along plate boundaries, and subsequent erosion have resulted in a vast array of island group shapes, most notably island arcs such as Japan or chains such as Hawai'i. The number and size of the islands in each island group vary, but more than 99% of the land mass is generally contained in islands >10 km² in area (Table 2.2). The smallest two island groups (Tonga and Canary Islands) are irregular clusters of nine to ten islands that are >10 km² in size, but of those, the Canary Islands are mostly >100 km², while all but two of the islands are <100 km². Tonga also has numerous islands <10 km². Jamaica and Hawai'i stand out as having no islands >10 and <100 km², while Puerto Rico and Iceland have very few islands >10 and <100 km². The largest three island groups have many islands in both size categories (Table 2.2). The two largest islands of New Zealand and the British Isles account for 99 and 66% of the total land mass of the island group, respectively, while the largest four islands of Japan, our largest island group, account for 96% of the total land mass. The total number of islands (typically an area >0.1 km²) for each group is an approximation as it depends on a consistent minimum size, sea-level status (small cays can be periodically inundated; Fig. 2.8; Box 2.4), and degree of exposure of land bridges. For the island groups, the number of islands >0.1 km² ranges from 4 in Jamaica, 9 to 17 in Puerto Rico, Canary Islands, and Iceland, 100–800 in Hawai'i, Tonga, and New Zealand, to >6000 in the British Isles and Japan. The number of inhabited islands increases as group area increases, but only Tonga, the British Isles, and Japan have substantial numbers of inhabited islands that are <10 km².

Fig. 2.8   Sooty terns nesting on Northeast Cay, Morant Cays, Jamaica; see Box 2.4.

Box 2.4. A day trip to the Morant Cays

Port Royal sits at the end of the Palisadoes, a long gravel spit that encloses most of the southern side of Kingston Harbor in Jamaica. It was the buccaneers' capital until its destruction in the earthquake of 1692. The rebuilt town did not fare much better in a more recent earthquake in 1907. Today it is a pleasant, small, and quiet town where it is good to go to eat fish and bammy cakes. It is also home to Jamaica's coast guard, and that is why I was there before dawn one morning in May, to leave on a trip on the coast guard boat to Jamaica's remote Morant Cays. These cays are tiny atolls that lie about 115 km east of Port Royal (and about 50 km from the easternmost tip of Jamaica), in the strait between Jamaica and Haiti. I was aboard on a trip taking staff from Jamaica's Natural Resources Conservation Department (NRCD) out to the atolls, where they were based for 2 or 3 weeks each year. The coast guard patrol boat was an open-topped boat with powerful engines and it sped rapidly into the slight chop of the warm Caribbean Sea along Jamaica's south coast, then out into open water. By late morning, the boat reached the cays. The biggest of the cays is about 300 m across and the cays are about 2 or 3 m above the sea. They are almost flat and fringed by reefs of the Morant Banks that stretch 15 km to the north and to the south of the cays, making them an important place for Jamaican fishermen to converge. Under the clear seas, the colorful corals provide habitat for many reef fish, but the sharks nearby made any prospect of snorkeling unappealing. On the cays there are a few low-growing, salt-tolerant plants but most of the area is covered with fine white sand. Over the cays were tens of thousands of nesting sooty terns and thousands of brown noddies nesting either in scrapes in the sand or in the little shelter that the low plants provide (Fig. 2.8). These seabirds were completely unafraid of people, barely moving away as we approached. The NRCD staff had come to conduct a census of breeding success and to ensure that the birds were left unmolested during their breeding season. The eggs fetch premium prices as delicacies, and some seabird populations on other cays have crashed because of overexploitation. The coast guard crew and I left the NRCD staff there to live under canvas tents for a week, and headed back to Kingston by dusk. PJB

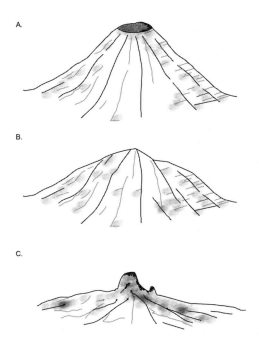

Fig. 2.9   Major stages in the erosion of a volcano. Redrawn from Whittaker and Fernández-Palacios (2007), by permission of Oxford University Press.

The spatial arrangement and topography of the islands in each island group reflect their geological history. Volcanic islands tend to be initially tall but erode over time forming steep-sided gullies (Figs. 2.9 and 2.10). The age of volcanic islands has been estimated from the degree of dissection. Limestone or coral islands and atolls are flatter and have lower elevations than volcanic islands; when they are uplifted they become known as makatea islands and have a rocky, coralline substrate. We now discuss the geography of each island group starting in Europe and going west.

2.3.1 British Isles

The British Isles (Fig. 2.11) extend 1230 km in a north–south direction and are geologically old and well-eroded, with the rugged highlands of Scotland, containing the islands' highest peak, Ben Nevis (1344 m). Several moors are found throughout the British Isles, while small hills are common in Wales and northern England, and broad river valleys

Fig. 2.10   Dissected, remnant crater wall from an ancient volcano on
the east shore of O'ahu, Hawai'i.

occur throughout much of southern England. There are many small
lakes and relatively short rivers. The largest lakes are Lough Neagh
in Northern Ireland (381 km²) and Loch Lomond in Scotland (71 km²).
The longest rivers are the Severn (354 km) and the Shannon (386 km).
Coastlines vary from steep limestone cliffs in southwestern England
and numerous bays and inlets on the western shores to generally
less dissected coastlines to the east and between Ireland and Great
Britain.

2.3.2 Iceland

Iceland (Fig. 2.12) stretches 525 km in an east–west direction and its
interior is a high elevation lava plateau punctuated by tuyas. Tuyas are
steep-sided mountains that were formed by volcanoes that erupted
under ice sheets. The volcano Hvannadalshnúkur, the highest point in
Iceland (2110 m) emerges from the Vatnajökull ice field, which is the
largest glacier in Europe (Plate 3). The southern coast has some flat
land used for grazing. Glaciers and lava fields each cover about 11%
of the island. There are many lakes, hot springs, geysers (an Icelandic
word), waterfalls, and rivers (the longest is the 230 km Thjórsá in
southern Iceland). The coastal topography includes wide glacial outwash plains and very abrupt cliffs such as the 440 m tall Látrabjarg

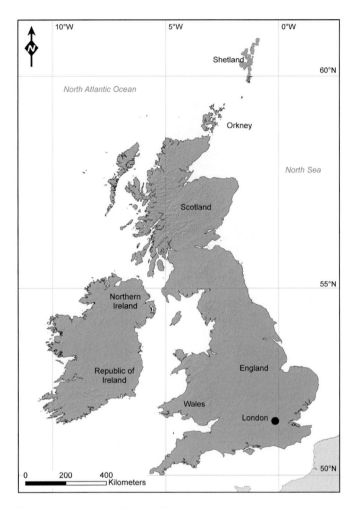

Fig. 2.11   Map of the British Isles.

cliff in far western Iceland that is a famous nesting site for some of Iceland's 3 million puffins and other seabirds (Plate 4). Despite extensive fjords in the west and northeast, Iceland's coastline is not particularly dissected compared to the other nine island groups, perhaps attesting to its youthful volcanic origin.

2.3.3 Canary Islands

The Canary Islands (Fig. 2.13) form a cluster of island along a 500 km long, east–west axis. Pico del Teide (3718 m), in the center of the

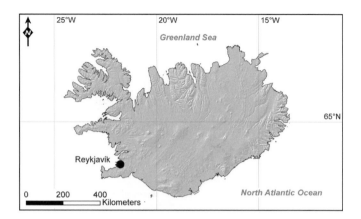

Fig. 2.12   Map of Iceland.

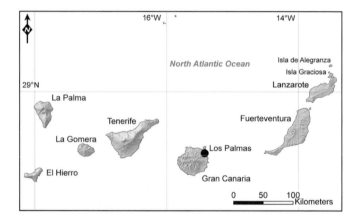

Fig. 2.13   Map of the Canary Islands.

largest island of Tenerife, is one of the world's largest island volcanoes and the highest point on the islands. Encircling the peak is
a giant, collapsed crater. The northern extension of Tenerife consists of old basaltic rocks that were probably a separate island at
one point but that are now connected due to past volcanic activity. In fact, the land area of the group has varied from its current
size to nearly twice that size during glacial advances due to volcanic
activity, subsequent erosion, and global sea-level changes. At one
point, about 13 000 years ago, Lanzarote and Fuerteventura were
joined into a single large island called Mahan. The interior of each
of the islands is a mixture of volcanic cones and craters, rugged

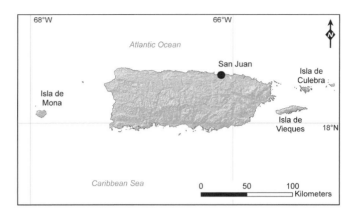

Fig. 2.14   Map of Puerto Rico.

lava fields (notably Montañas del Fuego on Lanzarote), and dunes on Gran Canaria. There are stream channels but few retain water year-round due to an arid climate and high levels of water use by humans (mostly for tourism and agriculture). Coastlines are highly dissected (Table 2.1) and characterized by cliffs, extensive beaches (Fuerteventura has 50 km of white sand beaches and 25 km of black sand), and dunes (Gran Canaria).

2.3.4 Puerto Rico

Puerto Rico (Fig. 2.14) extends 180 km in an east–west direction with a central mountain range dominated by the tallest peak (Cerro de Punta; 1328 m). The northwestern part of the island is dominated by the intricately eroded karst or limestone hills, with many caves, including some that streams flow through. There are no natural lakes (but many reservoirs), and many permanent streams flow down from the highlands to the coastal plains. The coast has alternating rocky and sandy shorelines and is not particularly dissected.

2.3.5 Jamaica

Jamaica (Fig. 2.15) extends 235 km in an east–west orientation. The Blue Mountains dominate the eastern end of the island and are the island's highest mountains, rising to 2256 m; they provide an abrupt topographical gradient to sea level (Plate 5). The central part of the remainder of the island is a 460 m tall limestone plateau called the

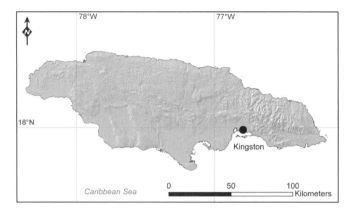

Fig. 2.15   Map of Jamaica.

Fig. 2.16   The Rio Grande drains the northeast slopes of the Blue
Mountains, Portland, Jamaica.

Cockpit Country because of the highly dissected karst topography
similar to that found in Puerto Rico: full of sinkholes, caves, disappear-
ing streams, and steep-sided hills. Surrounding this rugged terrain are
wide coastal plains, particularly to the west and south. Large interior valleys are traversed by many permanent streams (Fig. 2.16). The
northern coastline is characterized by uplifted coral reefs and white
sand beaches while the southern coast has limestone cliffs up to 300 m
tall and several black sand beaches. The most extensive beaches are on
the western end of Jamaica. Overall, the coastline is the third most dissected of the nine island groups, after Tonga and the Canary Islands.

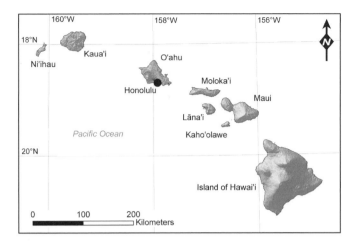

Fig. 2.17   Map of Hawai'i.

2.3.6 Hawai'i

The Hawaiian island chain stretches for nearly 3500 km across the central Pacific Ocean. The youngest island (the Island of Hawai'i; Fig. 2.17) has four peaks (Kohala, Hualālai, Mauna Loa, and Mauna Kea – the highest at 4207 m). Mauna Loa rises gradually from the ocean, has the characteristically flat profile of a shield volcano (see Fig. 2.3) topped by a summit caldera. The slopes of Mauna Kea and the much older Kohala are dotted with cinder cones. Loihi is the newest volcano in the Hawaiian chain but it may not rise above sea level for another 100 000 years. As older volcanic islands in the chain erode and subside, steep valleys are cut through the highly dissected slopes. Volcanic islands like Hawai'i gradually shrink in size until eventually the original caldera floor, hardened from multiple additions of lava, often forms an erosion-resistant plateau, which is the only remnant of the original shield volcano. The oldest volcanic remnants in the Hawaiian chain become seamounts as they erode below the surface of the ocean where they can form the base for the development of fringing coral reefs. There are permanent streams in Hawai'i; but the only lakes are small summit bogs and brackish coastal ponds. The only relatively level terrain occurs on eroded volcanic summits, in saddles between volcanoes or as gentle slopes near Hawaiian shore-lines. The relatively undissected coast of Hawai'i varies from steep cliffs such as occur on the northern shores of Moloka'i and Kaua'i and the eastern shore of O'ahu (Fig. 2.10), to the numerous but generally

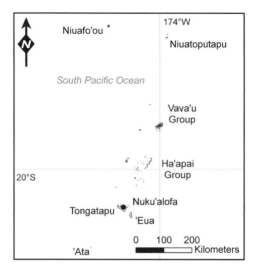

Fig. 2.18   Map of Tonga.

small beaches that have sands that come in various colors from white to black to olive green.

2.3.7 Tonga

The islands of Tonga occur in three main clusters (and several northern or southern outliers) that vary in age and origin (see Section 2.2). These clusters extend 750 km along the active north–south subduction zone of the Tonga Trench (Fig. 2.18). The highest point (1033 m) is on the volcanic Kao Island; about ten volcanoes currently rise above sea level, with some, like Fonuafo'ou, periodically submerging and re-emerging as a result of earthquakes, eruptions, and erosion. There are four islands with lakes and two with streams. Tonga has the most dissected coastline, relative to total island area, of any of the island groups as indicated by its high coast:area ratio (Table 2.1).

2.3.8 New Zealand

The New Zealand island group (Fig. 2.19) extends 3400 km along a north–northeast axis. The latitudinal range of New Zealand in the southern hemisphere approximates the range of Japan in the northern hemisphere (see Table 1.1). New Zealand's topography is dominated

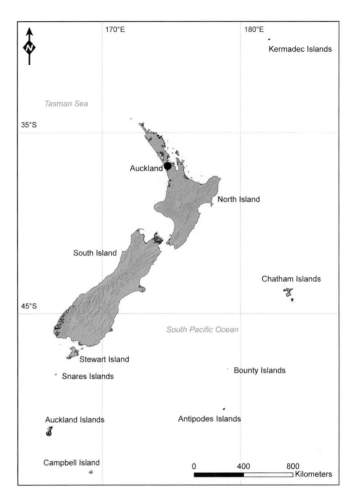

Fig. 2.19   Map of New Zealand.

by the mountains that have formed along the active plate boundary
that runs southwest to northeast along the axis of the South Island
creating the Southern Alps (Fig. 2.20), with the islands' highest peak
Aoraki (Mt. Cook; 3754 m). The fault extends northward and veers
east of the North Island. An arc of New Zealand's most recently active
volcanoes crosses the North Island and includes Taranaki, Ruapehu,
Tongariro, Ngauruhoe, Tarawera, and the youngest, Whakaari (White
Island), located 50 km off the northeast shore (see Fig. 3.2). The interior of New Zealand varies widely from the rugged Southern Alps
and northern volcanoes to wide coastal plains on the eastern side

Fig. 2.20   Southern Alps at the head of the Godley River, Mount Cook National Park, South Island, New Zealand.

of the South Island and rolling hills on much of the North Island. New Zealand has many rivers, lakes, waterfalls, and hot springs, with the largest lake (Taupo, 616 km²) and longest river (Waikato, 425 km) both on the North Island. Fjords characterize the southwestern coasts whereas cliffs and beaches are common throughout the country. Despite the fjords, the New Zealand coast is less dissected than most of the nine island groups.

2.3.9 Japan

Japan (Fig. 2.21) includes island arcs, circular volcanic islands, long north–south mountain ranges, and complex mixtures of tilted, folded, and eroded rock formations. The island group extends 3000 km in a northeast–southwest direction and about 75% of the land is mountainous with deep valleys cut by intensive erosion (especially on the eastern side), and also includes scattered plains, and inter-montane plateaus. Mountains on the western side are generally shorter and mixed with plateaus. The numerous volcanoes often form dramatic cones from successive and usually explosive eruptions (stratovolcanoes). Mt. Fuji is Japan's best-known volcano and also its highest peak (3776 m). The least mountainous island is Hokkaido, but even Hokkaido has eight distinct ranges of either hills or mountains, although several alluvial plains are also extensive. The largest island, Honshu, also has several coastal plains, including the largest Kanto

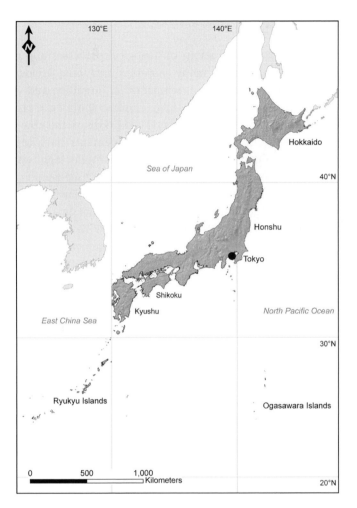

Fig. 2.21   Map of Japan.

Plain (13 000 km^2, where Tokyo is located), while the southern portion of the island is the most mountainous. Japan has numerous hot springs, lakes, and short, steep rivers. The largest lake (Biwa, 670 km^2) and the longest river (Shinano, 367 km) are both on Honshu. Japan's coastline is the longest but only fourth most dissected of the nine island groups yet is heavily indented with inlets, bays, and peninsulas. There are also many human modifications such as dikes and artificial islands that have been made to the coastline in order to increase habitable land area (see Section 2.2.5).

2.4 CLIMATE

Climate is a crucial determinant of biological processes and results from the combination of many variables including topography, oceanic influence, precipitation, temperature, humidity, and vegetation. Over geological time scales, huge climatic shifts have occurred across the Earth, so present-day climatic conditions on the nine island groups are transitory. Continental plate movements have certainly impacted island climates as islands shifted in latitude, longitude, and position relative to continents and ocean currents (see Section 2.2.2). Ice ages also affect island climates. For example, during the most recent Pleistocene glaciations (10 000–110 000 years ago), the British Isles and Iceland were entirely ice-covered, New Zealand and Japan were partly ice-covered, and climates varied from colder than at present (British Isles, Iceland), to generally cooler and wetter (Hawai'i and New Zealand), to cooler and perhaps drier (Japan and Tonga) to drier (Canary Islands). Interglacial periods were generally warmer and wetter than current climate conditions.

The nine island groups represent a wide range of precipitation and temperature conditions (Table 2.1) and fit seven of the nine major climatic zones used by Heinrich Walter and colleagues to describe world climates. The lowlands of Tonga, Puerto Rico, and Jamaica have an equatorial climate with more variation in daily than annual temperatures, and these islands have at least 100 mm of rain every month. At higher elevations, Puerto Rico and Jamaica have a tropical, summer-rainfall climate with greater seasonality in precipitation and temperature. Hawai'i has a wide range of climates from tropical dry and wet in the lowlands to more temperate conditions at higher elevations. The Canary Islands have a typical Mediterranean climate, with winter rainfall and long summer droughts. New Zealand, the British Isles, and southern and central Japan all have variations of a warm-temperate climate with a cool winter and warm to hot summers and a cool-temperate climate with cold winters and warm to hot summers, depending on the degree of oceanic influence. Iceland, northern Japan, and the northern British Isles have a cold-temperate climate with cold winters and cool summers. Iceland also has aspects of an arctic climate at higher elevations with long and very cold winters and short, cool summers. The two major climatic zones not extensively represented are subtropical dry and arid temperate climates, although parts of most of the island groups are arid.

2.4.1 Topography

Islands with mountains are generally wetter than islands with flat-ter terrain because in the presence of offshore winds (winds from the ocean), island mountains force warm, moist air to rise, expand, and cool, thus reducing its capacity to hold moisture. The result is rain (orographic precipitation) on the windward side of the mountains (Fig. 2.22). Windward places in Hawai'i receive up to ten times the rain-fall found over the nearby open ocean because of this phenomenon. Where strong, steady trade winds continue to push the air up, then over and down the other side of a mountain (Fig. 2.23), the descending and warming air absorbs moisture from the land, creating the dry lee-ward conditions of a typical rain shadow (Fig. 2.24). Tonga, the Canary Islands, Puerto Rico, Jamaica, and Hawai'i are in the trade wind zone where winds cross the oceans and reach these mountainous islands from the northeast (in the northern hemisphere). Consequently, the northeastern sides of mountains in the Canary Islands, Puerto Rico, Jamaica, and Hawai'i experience wetter climates than the south-western sides. The plants respond to these changes in rainfall and there are typically well-defined elevation zones of vegetation, includ-ing cloud forests that occur in the zone of maximum precipitation.

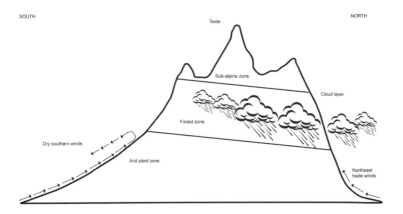

Fig. 2.22 Rainshadow on mountainous islands in trade-wind zones such as found on Pico del Teide in the Canary Islands. The windward side receives substantial rain because moisture-laden winds from the ocean rise and then lose their moisture as rain on the upper slopes. The leeward side is dried by cool winds warming and absorbing moisture as they come down the other side of the mountains. Redrawn from Whittaker and Fernández-Palacios (2007), by permission of Oxford University Press.

Fig. 2.23   Northeasterly storm clouds blown over the Grand Ridge of
the Blue Mountains, Jamaica by trade winds into the Green River Valley.

Interestingly, the vegetation also influences the elevation at which
rain becomes common. Following extensive defoliation of forests on
windward Puerto Rico after Hurricane Hugo in 1989, the lower elevational limit of frequent cloud cover rose because there was less transpiration (loss of water) from the defoliated trees (a phenomenon not
limited to island climates).

Climatic patterns become more complex on the larger island
groups but remain strongly influenced by topography. In Iceland,
mountains contribute a 40% increase in rainfall compared to a hypo-
thetically flat island. Iceland has a pronounced rain shadow on the
north (leeward) side of the island. New Zealand's winds and rains
come primarily from the northwest, and on the more mountainous
South Island there is a sharp precipitation gradient from the very wet,
west coast to the drier inlands on the eastern side of the Southern
Alps. Winds in the British Isles generally come from the southwest
and the wettest parts of the islands include the mountains in Wales,
Scotland, and the Pennines in central England. Japan has strongly
seasonal weather and the wet summers come from the interception
by its many mountain ranges of moist tropical monsoon winds from
Southeast Asia. Winters in Japan are dominated by cold air flowing
eastward from Siberia that mixes with water vapor from the relatively
warm Sea of Japan to cause snow on the windward side and a "snow
shadow" on the leeward side.

Fig. 2.24   (A) Lowland rainforest on limestone, eastern slopes of the
John Crow Mountains, on the windward side of Jamaica. (B) Dry forest
on limestone, Calabash Bay, on the leeward side of Jamaica.

2.4.2 Oceanicity

Oceanicity encompasses the total effect of the ocean on island (or coastal) climates and is affected by length of coastline, distance to mainland, island size, ocean currents, and accompanying temperature and humidity conditions. Because of their extensive coastlines, the nine island groups have oceanic climates, even when they are relatively large and close to a mainland such as is the case for the British Isles. Oceanic climates are more humid than continental ones that are distant from a coastline (such as in central Canada). The greater humidity of oceanic climates buffers temperature extremes, resulting in cooler summers and warmer winters than continental climates found inland. The temperature of the prevailing ocean current also affects climate, with cool currents providing little moisture to the surrounding air (associated with dry coastlines and nearby islands such as the Canary Current that passes the Canary Islands). In contrast, warm currents provide abundant moisture to the islands they encounter (such as the Gulf Stream that flows past Jamaica, Puerto Rico, Iceland, and finally reaches the British Isles).

2.4.3 Precipitation

The amount of precipitation that falls on an island is a crucial factor in determining what plants and animals live there. The Canary Islands are overall the driest of the nine island groups (Table 2.1) while New Zealand, Hawai'i, and Japan are the wettest. Indeed, several locations argue that they receive the most rainfall in the world (including a bog on the top of Kaua'i), but that accolade can be measured in so many ways, including total annual rainfall, hours of rainfall, highest daily mean rainfall, or fewest numbers of days without rainfall, that the argument remains unresolved. In addition, such wet places tend to be inaccessible and difficult to collect data in (because rain gauges overflow or electronic sensors short out). Precipitation comes largely as rainfall but also as snow, fog, hail, or sleet. Three other forms of suspended precipitation that can influence climate include salt spray, virga (rain that does not reach the ground), and vog (volcanic fog that forms when sulfur dioxide gas mixes with sunlight, oxygen, and dust particles in the air). Vog can burn the leaves of plants, harm humans with respiratory illnesses, and cause acid rain. Since 2008, there have been renewed problems with vog on the Island of Hawai'i related to new eruptive activities on Kīlauea Volcano.

Broad spatial patterns of rainfall reflect topographical gradients and ocean currents but also may have the seasonal component noted above. The Atlantic island groups (Iceland, the Canary Islands, and the British Isles), tend to be driest during May to August (northern hemisphere summer); this pattern is most pronounced on the eastern coast of the British Isles. The tropical islands of Tonga, Puerto Rico, and Jamaica have little seasonality in precipitation but there is a short winter dry spell from July to September (Tonga) or January to March (Puerto Rico and Jamaica) that is sometimes extended into the rest of the year on the drier southern slopes of Puerto Rico. Japan has the most widespread seasonal consistency in precipitation among the larger islands, with the wettest part of the year from May to August or sometimes later in the year to the north. The abrupt seasonal shift from Siberian cold to monsoonal warm and wet gives Japan its predictable spring cherry blossoms. Hawai'i's precipitation is highly seasonal (except in the wettest and driest portions of the islands), and has the least precipitation from May to August. The western coasts of New Zealand are generally wettest during November to February, and the eastern coasts are driest during that same period. Again, as with temperature, variability is the norm so that New Zealand's weather forecasters optimistically talk of the weather "becoming fine" but are never sure what to expect on any given day. Local features also impact precipitation. These can include local vegetation that tends to add moisture through evapotranspiration or small-scale topographical features, such as hills that deflect either dry or moist winds.

2.4.4 Temperature

Temperature is another variable that has critical effects on life; it is affected by seasonality, topography, latitude, ocean currents, cloudiness, vegetation, and precipitation, among other variables. Mean annual temperatures on the nine island groups vary from the clearly tropical climates of Puerto Rico and Jamaica to the cold-temperate climate of Iceland. Seasonal fluctuations in temperature are minimal in the tropical regions for Tonga, Puerto Rico, and Jamaica with typically no more than several degrees difference between winter and summer mean temperatures. However, larger temperature ranges for Hawai'i (Table 2.1) reflect the intra-island variation by elevation, not a stronger seasonal variation at any site. A drop in annual temperature of 0.5°C is typical for every 100 m elevation gain in Hawai'i. High elevation sites such as the volcanic summits of Mauna Kea and Mauna Loa on the

Island of Hawai'i and Haleakalā on Maui are much cooler (the mean annual temperature on the summit of Mauna Loa is 5°C) and have slightly more seasonal change in temperatures than the Hawaiian lowlands. The cooling effect of increased elevation can be found on all the nine island groups. Iceland, New Zealand, the British Isles, and Japan all show a strong seasonal pattern of warm summers and cool to cold winters, with the strongest seasonal variation in Japan. The Canary Islands are also seasonal but asymmetrically so, with a faster warming period in May and June than the cooling period in September through November. On most of the island groups, humans have directly altered island temperatures by the creation of urban heat islands and indirectly through deforestation that has resulted in less cloud cover, less precipitation, and therefore higher temperatures and a drier climate. In the next section, we discuss how temperature and precipitation interact.

2.4.5 Climate measurement and integration

How climatic variables are measured depends on the purpose of the measurement and the spatial and temporal scales of interest. Details about conditions in the immediate area (microclimate; about $1-10\,m^2$) are relevant for plant physiology or animal behavior; mesoclimate refers to area scales of $10-1000\,m^2$ that are relevant to animal population dynamics; while macroclimate encompasses entire landscapes ($>1000\,m^2$) where one can observe effects of spatial gradients on organisms. Each of these spatial scales requires a different approach to climate measurements. Climate variables such as precipitation and temperature can also be measured at many temporal scales to explore variation by minute, day, month, year, or longer. One can then summarize climate variables during any given time frame by calculating means, ranges from minimum to maximum, rolling averages, or even determine how many days are above or below a certain cutoff value of temperature or precipitation (useful to gardeners, for example, to determine growing seasons). However, generalizing island means or annual means of precipitation or temperature obscures the extraordinary variability in spatial and temporal climatic conditions.

Precipitation and temperature are linked through the process of evapotranspiration, the transfer of water from the surface of the ground and from organisms into the air. The potential evapotranspiration or PET increases with temperature but actual evapotranspiration (AET) is highest in hot, wet climates. AET is directly linked to plant productivity, so the hot, humid parts of tropical islands, such as Puerto Rico and Jamaica, have high AET and dense forest growth. AET is lowest on the Canary Islands, the driest of the nine island groups, despite the high annual temperatures because there is little moisture to evaporate. Additionally, because of this close interaction between precipitation and temperature, the same amount of precipitation (such as 1000–1500 mm per year) can mean moist to wet conditions in a cool climate and almost dry conditions in a hot climate. Humidity is both a result of AET and an influence on it. At a given temperature, increases in humidity reduce AET because there is less difference between the moist surface and the wet air than under drier conditions. AET is therefore a valuable tool to help interpret the influence of both precipitation and temperature on organisms.

2.5 SOILS

Soil formation is a function of the geological substrate interacting with climatic factors, topographical features, plants and animals, and time (Box 2.5). Development of fertile, humus-rich soils can take thousands to even millions of years, but is more rapid in warm, humid environments and where there is ample plant growth. Soil loss occurs from erosion (see Chapter 3) by wind and water and disrupts the horizontal layers of soil that have slowly developed. Rapid geological uplift on several island groups, such as New Zealand and Japan, triggers equally rapid down slope movement through erosion (Fig. 2.25).

Box 2.5. Island contributions to science: how do soils form?

The major islands of Hawai'i form a 4.1 million year time series from the youngest (Island of Hawai'i) to the oldest (Kaua'i) that has been very helpful for studying how soils develop. Newly deposited Hawaiian lava is relatively rich in some critical nutrients such as phosphorus but lacks nitrogen. The first organisms on lava are usually windblown spiders and crickets that often survive by eating each other. Plant material also blows in and soon plants begin to grow in pockets where organic debris has accumulated. Eventually forests develop except in very dry parts of the islands. I participated in studies on the Island of Hawai'i that confirmed reports from New Zealand that a lack of adequate soil nitrogen limits plant growth during the first several thousand years of soil

Box 2.5. (cont.)

development. Later, more extensive studies found that phosphorus continued to decrease for millions of years of soil development because phosphorus leaches and becomes immobilized over time. Loss of phosphorus is partially offset by inputs in dust deposits blowing in from Asia. Other studies of soil formation include work in Iceland on lava flows and in Puerto Rico on landslides, highlighting the importance of islands to science. LRW

Fig. 2.25   A landslide in Puerto Rico where soil and vegetation loss initiate the processes of new soil development and plant community recovery.

2.5.1 Geological substrate

There are five extensive geological substrates upon which soils develop on the nine island groups. The first and most widespread are substrates of volcanic origin. Volcanoes produce four main forms of lava (basalt, andesite, rhyolite, and dacite) that vary in their viscosity, content of silicon dioxide, and other chemical compounds. Volcanoes also eject material into the air that is collectively termed tephra (including ash, cinder, and pumice). Volcanic substrates can support development of all of the twelve soil orders of the world, depending on the climatic conditions. Ten of these soil orders occur in Hawai'i alone, although over half of Hawai'i's land surface has no soil development because it remains lava, tephra, coral outcrops, beaches, and land created by erosional processes filling in shallow coastal areas. On lava or tephra surfaces that are relatively new or unstable, soil formation is just beginning. Poorly developed soils like these also occur in Iceland, New Zealand, Tonga, and Japan. More time and weathering can produce fertile soil types in wet areas with poor drainage (such as on ridget-ops of Japan or Hawai'i) and also in better drained areas. Less fertile soils develop in the more arid zones of the Canary Islands. Soils found in colder climates as on Iceland are intermediate in fertility and are highly organic but many nutrients have been leached out; Icelandic soils are also typically underlain by permafrost.

Limestone is another common substrate on the island groups. It comes from uplifted coral reefs or old sea beds and can be fairly pure calcium carbonate, such as the chalk cliffs of southwestern England, or mixed with various clays, sands, iron oxide, or other materials. Limestone is more resistant to erosion than other sedimentary rocks so it often forms remnant ridges and outcrops that can be quite highly dissected (karst topography). Most caves occur in limestone where water has gradually dissolved the limestone and drained away through increasingly wider cracks. The porous nature of limestone means that surface water (lakes and rivers) are uncommon in the overlying sub-strate. Soils that develop on limestone substrates vary but are usually high in calcium. The limestone-derived soils in Jamaica are weathered red and brown soils that are low in other bases, high in aluminum, poorly aerated, and with low water-holding capacity.

A third common geological substrate called greywacke is a type of sandstone that forms the basis for soils over about half of New Zealand and is abundant in the mountains of Ireland, Wales, Scotland, and northern England. Greywacke is formed from the mixing of sand and mud on an ocean floor that then gets mixed with quartz and feld-spar and undergoes deformation and hardening; it is very nutrient-poor.

A fourth major source of soils on the island groups is from allu-vial processes where sediment is deposited, such as along the banks of rivers. Many of the island groups such as Jamaica, Puerto Rico, New Zealand, and Japan have large coastal plains formed by alluvium brought down from central highlands through water erosion and sediment transport. Alluvial soils can be quite fertile and are often used for agriculture including sugar cane production in Puerto Rico, Jamaica, and Hawai'i and rice cultivation in Japan.

A fifth substrate for soil development is aeolian deposits that are sediment transported by wind. Particularly fine wind-blown sands or clays called loess can form deposits tens of meters deep (as in Japan). Central Iceland has a particularly rapid accretion of aeolian substrates of one or more centimeters per year on the tops of remnant hills, called rofabards while the sides erode, often to bedrock, leaving sharp cliff edges.

Soils often form from various combinations of these five substrates. For example, soils in Tonga frequently form from layers of limestone and volcanic ash. Loess from Central Asia and the Sahara Desert has mixed with locally derived volcanic material to form soils in Hawai'i and the Canary Islands, respectively. There are many other less extensive geological substrates present including granitic intrusions (as found in Puerto Rico and Jamaica), gneiss (northwestern Scotland), slate (Wales), ultramafics (northern British Isles and New Zealand), and sandstone (North Island of New Zealand), each with its unique influence on soil development. Such a variety of soil types supports the observation that there are no two soils that are exactly alike. The addition of topographical, climatic, and biological variability further expands the range of possible outcomes for soil development.

2.5.2 Topographical influences

Topography plays a crucial role in determining what types of soils develop across a landscape, particularly as it affects elevation, slope, and drainage. The influences of elevation on soil development usually involve changes in precipitation and temperature due to changes in elevation. Low-elevation, coastal soils can also be modified by salt spray, often reducing their development and fertility but also adding critical nutrients. Steep slopes are less stable than flatter terrain, so soil development on slopes is delayed until a degree of stability is achieved following extensive erosion. The degree of drainage is also a key factor in determining soil type. Soils that are rich in organic matter often form where drainage is moderate. Poor drainage leads to less soil development because of less decomposition, lower levels of plant-available nutrients, poor aeration, and slow nutrient recycling by soil microbes. Better drainage usually means more microbial activity and more available nutrients but high drainage conditions can create highly leached soils where nutrients are removed. The recognition of the importance of topography has led to soil classifications based on slope and drainage. For example, in Puerto Rico, the US Soil Conservation Service recognizes five different soils based in part on topographical position.

2.5.3 Climatic influences

Precipitation and temperature are both critical to soil development and their influences are closely intertwined. Extremes, normal ranges, and averages of precipitation and temperature each has its impact on soil development, as do amount and timing of precipitation. Extreme heat, drought, and cold are not favorable for most soil organisms and therefore to soil development. Torrential rains are normal for many biomes, but sudden, unusually heavy precipitation can cause soil erosion in any climate. Soil development is driven by the sum of all geographical, topographical, climatic, and biological influences. For example, in Jamaica, the Blue Mountains are rainy and cool, and receive the bulk of the precipitation from trade winds, but have poor soil development because of infertile rocks. On the alluvial plains of the southern leeward side of the island, precipitation is lower than in the mountains, but the soils are more fertile and better developed.

2.5.4 Biological influences

Soil organisms construct soil over time provided that the physical conditions are favorable for their metabolism. They provide the role of decomposer, recycling the complex carbohydrates and lignins formed by plants. Plants then compete with the soil microbes for the nutrients that are released, even playing an active role in manipulating their immediate soil environment to favor their own growth over that of other organisms. Some plants have a particularly strong influence on soil nutrient availability because they have the ability to capture atmospheric nitrogen and convert it to forms usable for microbes and plants. The aboveground parts of plants also influence soil development by cooling soil temperatures, stabilizing the substrate, and adding nutrients through leaf litter. Plants sustain animals that, in turn, alter soil development because they aerate, burrow, defecate, and die in or on the soil. The quality of plant litter (dead leaves, stems, roots, fruits, seeds) has a major influence on soil texture and nutrient status. For example, the slow-growing, evergreen conifers in New Zealand provide a low-nutrient leaf litter that is slow to decompose and results in low-nutrient soils.

2.6 SUMMARY

Islands are not unlike mainlands: they have mountains, lakes, flood-plains, and other familiar features. While volcanism is more wide-spread among the island groups compared to all terrestrial surfaces, most of the world's soil orders are represented on the nine groups. This highlights how factors other than geology are also crucial to soil development. The climates of the island groups are, of course, more oceanic than average. In addition, their geographical isolation and limited size influence biological invasion and evolution in unconventional ways. We use the description of the physical aspects of these islands developed in this chapter to understand the dynamics of those physical parameters in Chapter 3, where we focus on the present-day disturbance regime not impacted by humans. Then, in later chapters, we address how the geography of these islands influences plant and animal invasions and how humans arrive, disrupt, and otherwise respond to these ecosystems.

SELECTED READING

Agricultural Research Institute of Iceland (2005). Icelandic soils. http://www.rala.is/desert/2~1.html (accessed 24 June 2009).

Bardgett, R. (2005). *The Biology of Soil: A Community and Ecosystem Approach.* Oxford: Oxford University Press.

Bouysse, P. (1988). Opening of the Grenada back-arc Basin and evolution of the Caribbean plate during the Mesozoic and Early Paleogene. *Tectonophysics* **149**: 121–143.

Buskirk, R.E. (1985). Zoogeographic patterns and tectonic history of Jamaica and the northern Caribbean. *Journal of Biogeography* **12**: 445–461.

GLGARCS (2009). http://www.glgarcs.net (accessed 24 April 2009).

Graham, A. (2003). Geohistory models and Cenozoic paleoenvironments of the Caribbean region. *Systematic Botany* **28**: 378–386.

Karan, P.P. (2005). *Japan in the 21st Century: Environment, Economy and Society.* Lexington, KY: University Press of Kentucky.

Kay, E.A. (ed.) (1994). *A Natural History of the Hawaiian Islands. Selected Readings II.* Honolulu, HI: University of Hawai'i Press.

Nunn, P.D. (1994). *Oceanic Islands.* Oxford: Blackwell.

Pilkey, O.H. and M.E. Fraser (2003). *A Celebration of the World's Barrier Islands.* New York: Columbia University Press.

Surtsey Research (2009). *Surtsey Research Volume 12.* Reykjavík: The Surtsey Research Society. http://www.surtsey.is (accessed 22 October 2010).

Taylor, B. (ed.) (1995). *Backarc Basins: Tectonics and Magmatism.* New York: Springer.

Te Ara Encyclopedia of New Zealand (2009). http://www.teara.govt.nz/
EarthSeaAndSky/Geology (accessed 20 June 2009).

Terrestrial Ecoregions (2009). Canary Islands dry woodlands and forests. http://
www.worldwildlife.org/wildworld/profiles/terrestrial/pa/pa1203_full.html
(accessed 20 June 2009).

University of Idaho (2009). Soil orders. http://soils.ag.uidaho.edu/soilorders/
orders.htm (accessed 24 June 2009).

Vitousek, P. (2004). *Nutrient Cycling and Limitation: Hawai'i as a Model System.*
Princeton, NJ: Princeton University Press.

Wagner, W.L. and V.A. Funk (eds.) (1995). *Hawaiian Biogeography: Evolution on a Hot
Spot Archipelago.* Washington, DC: Smithsonian Institution Press.

Walker, G.P.L. (1990). Geology and volcanology of the Hawaiian Islands. *Pacific
Science,* **44**: 315–347.

Walter, H., E. Harnickell and D. Mueller-Dombois (1975). *Climate-diagram Maps
of the Individual Continents and the Ecological Climatic Regions of the Earth.* Berlin:
Springer.

Whittaker, R.J. and J.M. Fernández-Palacios (2007). *Island Biogeography: Ecology,
Evolution, and Conservation.* 2nd edn. Oxford: Oxford University Press.

3

Natural disturbances on islands

3.1 DISTURBANCE CHARACTERISTICS

The geological and geographical processes discussed in Chapter 2 are dynamic over millions of years as islands form and change shape. Changes in the physical environment also occur over annual and decadal scales that are more relevant to plants and animals on islands. When these changes disrupt living organisms by causing a reduction of biomass (plant or animal matter) or structure (such as the canopy of a forest) we call them disturbances. Some of these disturbances are on-going tectonic events, such as volcanoes, earthquakes, and slope erosion. Other disturbances arise from the presence of the organisms themselves, such as when seabirds dig burrows in the ground or flammable plants promote the spread of fires. In this chapter, we discuss both the physical and the biological disruptions to the plants and animals on the nine island groups. The emphasis in this chapter is on natural disturbances as we set the stage for understanding the arrival and evolution of the diversity of plants and animals that inhabit islands (see Chapter 4). These initial chapters then provide a framework within which to understand the influences of humans on island biota (anthropogenic disturbances, see Chapters 5–8).

Disturbances that damage living organisms are relatively discrete events that alter the structure of a community or ecosystem. Events that occur over longer periods of time, such as a gradual depletion of nutrients from the soil or a gradual change in air temperature are considered stresses rather than disturbances. Disturbances can occur at many spatial scales, from local to global. At a local scale, a leaf can be damaged by a foraging insect. At a global scale, volcanic ash eruptions can cause a sudden drop in temperatures by reducing sunlight, as occurred in 1991 with the eruption of Mount Pinatubo in the Philippines. The global cooling that resulted from that eruption caused reductions of seed production by native beech trees in New Zealand forests. Therefore, what comprises a disturbance depends on the spatial and temporal scale of interest. Disturbances can be viewed from the nature of the physical force, from the effects on organisms, and from the response of the organisms.

The nature of the physical force that causes a disturbance varies in intensity (such as wind speed of a cyclone), extent (the area covered by the wind), and frequency (such as mean cyclone return interval for a given plot of land). Each disturbance also has its own particular traits, including local turbulence within a cyclone. The effect of any given disturbance is measured by its severity or the amount of direct damage it causes (such as percent of trees that it knocks over). Intensity and severity are considered aspects of the overall magnitude of the event. There are also many indirect and long-term effects that become apparent to careful observers of disturbances. These effects influence not only local organisms but can also alter topographical features and even local climate. Finally, the response of organisms to the disturbance is complex, with both short- and long-term components. One well-studied aspect of disturbance responses is the process of ecological succession, the changes in species composition and ecosystem attributes that take place after a disturbance. The presence or absence of residual organic matter and soil after the disturbance determines the speed of the recovery process for plants and animals.

Each disturbance, with its characteristic intensity, extent, frequency, and severity is part of a larger set of disturbances called the disturbance regime that affects a particular site. These disturbances interact and potentially influence each other, as when earthquakes trigger landslides and landslides dam the flow of rivers. The feedback from the disturbed organisms can also sometimes modify the physical aspects of the disturbance regime. We will elaborate on some of the characteristics of individual types of disturbance in later sections.

We can group disturbances by the Greek elements (earth, air, water, fire) that they primarily involve. This approach focuses on the main physical driving force and is a useful way to categorize disturbances. In this book, earth represents tectonic disturbances including volcanoes, earthquakes, erosion (landslides), and land building (dunes). Air includes cyclones. Water includes floods (from glaciers, rivers, and along coastlines), drought, and ocean impacts (tsunamis). Fire is a common natural disturbance on many continents, but natural fires are less common on most islands. A fifth type of disturbance is caused by the action of animals, as when locust swarms or other insect outbreaks damage the vegetation.

Another classification of disturbances uses the degree of severity, so that both a cyclone and a volcano that destroyed 85% of all local organisms would be in the same category. This approach allows comparisons across disturbance types and is useful when the focus is on severity. A third approach classifies disturbances by the type of surface that they leave behind, such as smooth lava or a silty floodplain. This approach is useful when the interest is in subsequent processes of recovery through succession. Surfaces that are left unstable and infertile, such as the slip face of a landslide, will be slow to recover, while surfaces that are left stable and fertile, such as a plowed agricultural field, will recover more quickly. Other factors that influence the rate of recovery include the climate, geology, and surrounding vegetation and whether the vegetation is a sources of potential colonists for the newly formed habitat.

The nine island groups represent an intriguing cross-section of disturbance regimes with many parallels to mainlands. Certainly, none of the disturbances discussed here is unique to islands. However, in combination, the physical aspects of islands (isolation, finiteness) and the biological response to island disturbances as a result of their remoteness make islands particularly vulnerable to disturbances. Natural disturbances that are most prominent on islands include volcanoes, earthquakes, cyclones, and disruption by seabirds. Many islands are formed from volcanoes and continue to be disrupted by tectonic activity. Tropical cyclones are generated by warm ocean water and tropical islands are frequently in the path of these cyclones. Seabirds prefer to breed and nest on isolated islands and can have major impacts on vegetation and nutrient cycles. Disturbances less important to islands than to mainlands include drought, flooding, and fire, although these disturbances are not unknown to islands, particularly if the landscape has been altered by humans. We describe the general characteristics of each disturbance, illustrated by representative examples from the island groups. Then we discuss the ecological effects of and biological responses to the disturbance.

3.2 VOLCANOES

3.2.1 Characteristics and examples

Volcanism is one of the most obvious and disruptive disturbances on islands and has been a major part of the long-term history of each of the nine island groups. Volcanism remains an important disturbance on all the island groups except Jamaica, Puerto Rico, and the British Isles. For most visitors and some residents, the potential for flowing lava or erupting ash adds a dash of danger to the mystique of islands. The four oceanic island groups (Tonga, Canary Islands, Hawai'i, and Iceland) owe their existence entirely to volcanoes and still are volcanically active. The British Isles are part of a continental shelf where the last volcanism occurred about 55–60 million years ago (see Chapter 2). Puerto Rico and Jamaica are continental fragments where volcanism ceased at least 17 million years ago. In contrast, New Zealand and Japan are continental fragments that are still volcanically active.

Volcanoes are caused by fissures in the Earth's crust that permit the release of hot molten rock, ash, and gases to the Earth's surface. The intensity of a volcano is often measured by the Volcanic Eruption Intensity (VEI) scale that goes from 0 (no explosive action; Mauna Loa, Hawai'i) to 9 (Table 3.1) and increases by a factor of 10 with each higher level. Taupo in New Zealand reached a VEI value of 7 in AD 181 in an eruption that ejected 120–150 km³ of material in a column reaching 50 km into the atmosphere. This eruption was big enough to be documented by Roman and Chinese scholars, who noted the unusually colored sunsets that resulted from the ash. However, this eruption, the largest on Earth during the last 2000 years, was dwarfed by an earlier eruption at Taupo 26 000 years ago (VEI = 8), that ejected 1000 km³ of ash. The extent of volcanic eruptions depends on their intensity and also on factors such as wind conditions that determine the spread of ash. Very large volcanoes can cool global temperatures because the ash gets into high elevation wind currents and reduces sunlight around the world. More often, volcanoes impact local areas downwind of the eruption that encompass tens of square kilometers. In 2010, Eyjafjallajökull Volcano in Iceland severely curtailed air traffic in Europe for weeks because of the danger high elevation ash presented to airplane engines.

About 75% of the world's active volcanoes are part of the Pacific Ring of Fire, a zone of converging plates on which New Zealand, Tonga, and Japan are located. New Zealand has several active volcanoes that

Table 3.1. *Volcanic Explosivity Index (VEI) and characteristic volcanoes. Each increase is equivalent to an eruption 10-fold more intense. High viscosity lava is associated with more intense eruptions and steeper cones. CI = Canary Islands, HI = Hawai'i, IC = Iceland, JP = Japan, NZ = New Zealand; dates are AD unless otherwise noted; My = million years ago*

VEI (magnitude)	Description	Plume height (km)	Volume	Frequency	Examples
0	Non-explosive	0	>1000	Daily	Mauna Loa (HI) 1984
1	Gentle	0.1–1	10000 m^3	Daily	Kīlauea (HI) 1983–present
2	Explosive	1–5	1 million m^3	Weekly	Usu (JP) 2000–1; White Island (Whakaari) (NZ) 2001
3	Severe	3–15	10 million m^3	Yearly	Surtsey (IC) 1963-7; Eldfell (IC) 1973; Ruapehu (NZ) 1995-6
4	Cataclysmic	10–25	>0.1 km^3	Decadal	Hekla (IC) 1158, 1300, 1510, 1693
5	Pyroxysmal	>25	>1 km^3	Centennial	Teide (CI) 0.15 My; Edgecumbe (NZ) 300; Fuji (JP) 1707; Askja (IC) 1875; Tarawera (NZ) 1886
6	Colossal	>25	>10 km^3	Centennial	Eldgjá (IC) 934; Laki (IC) 1783
7	Supercolossal	>25	>100 km^3	Millennial	Taupo (Hatepe; NZ) 181; Kikai (JP) 4 300 BC
8	Megacolossal	>25	>1000 km^3	Every 10000 years	Taupo (Oruanui; NZ) 24500 BC
9	Gigacolossal	>25	>10000 km^3	Very rare	La Garita (Colorado, USA) 26–28 My

have erupted in the last century, including Ngauruhoe (1977), Ruapehu (1996), and Whakaari (White Island) that has nearly continuous gas emissions interspersed with small eruptions. Fourteen volcanoes have erupted in Tonga in the last 10 000 years, including seven that are now submerged. In 2006, an eruption of one volcano near Late Island created a belt of densely packed pumice many kilometers wide that rafted across the sea, slowing the progress of a yacht that sailed into this "sea of stone." The latest to erupt, Hunga Tonga–Hunga Ha'apai, is located 10 km from Tongatapu. Japan has more than 80 active volcanoes, or about 10% of the world total despite having 1/400 of the Earth's land surface. Most are on Honshu (47) followed by Hokkaido (18) and Kyushu (9) with the remainder on the Ryukyu Islands (7) and other small islands such as Iwo-jima. The eruption of Kikai in the northern Ryukyu Islands 6300 years ago had a VEI of 7, making it one of the more powerful eruptions in the last 10 000 years. Ash from its eruptions covered most of Japan and made Kyushu uninhabitable by humans for several hundred years. Today, its 19 km wide caldera is mostly submerged, but small islands are being formed around the rim, which have erupted as recently as 2004.

Volcanoes are also found where tectonic plates pull apart such as on the mid-Atlantic Ridge where Iceland is located. Iceland rivals Japan as a global hotspot of volcanic activity and has some fascinating volcanoes. Eighteen volcanoes have erupted since human settlement in AD 874 and 11 have erupted in the last 100 years. During the last 500 years, Iceland's volcanoes have produced a third of global lava output. Eldgjá features the largest volcanic canyon in the world (270 m deep, 600 m wide) and in 934 produced the largest lava flow in a single eruption in recorded history (19.6 km^3). The nearby Laki Volcano erupted in 1783/4 and produced a 14 km^3 flow of lava and clouds of poisonous gas that killed 50% of Icelandic livestock, which resulted in a famine that ultimately killed 25% of Iceland's population. A 1973 eruption on Heimaey, an island off the southern coast of Iceland, nearly closed the only harbor on the island (Fig. 3.1), but was thwarted by the heroic efforts of local citizens who sprayed sea water from fire boats.

The islands of Hawai'i have been created during the last 80 million years over a hotspot or mantle plume that is not associated with a plate boundary. Current volcanic activity is limited to the two youngest islands. On Maui, Haleakalā erupted most recently in 1790. On the Island of Hawai'i, Mauna Loa has erupted 38 times in the last 200 years. It is an example of a shield volcano with its flattened profile formed from low-viscosity magma that readily flows long distances. Recent

Fig. 3.1   The town of Heimaey, Iceland (background) that saved its
harbor from closure by a volcano by spraying sea water from fire boats
onto the expanding lava (foreground) of a 1973 volcano.

activity has been from Kīlauea, which has erupted continuously since
1983 (see Box 2.1). In 2008, the Halemaʻumaʻu summit crater of Kīlauea
erupted explosively for the first time in several decades.

The Canary Island volcanoes have been active for about 20 million years with recent activity on five of the islands and examples of
both shield volcanoes (Gran Canaria) and stratovolcanoes (Tenerife). The
youngest volcano in the Canary Islands is Cumbre Vieja that erupted
most recently in 1949 and 1971. Many dramatic collapses of the steep-
sided volcanoes of the Canary Islands and extensive erosion have produced a rugged and varied terrain of fissures, vents, and gorges.

3.2.2 Ecological effects and responses

The response of organisms to a volcanic eruption depends on the severity and extent of the disturbance. Recovery on lava is generally very
slow because of the nearly total absence of organic matter, although

cracks in smooth (pahoehoe) lava surfaces can fill quickly with wind-blown sediments that then support plant growth if the environment is sufficiently moist. Rough or clinker lava (aʻa) is slower to be colonized because it is generally better drained and does not as readily collect sediments or moisture. The depth of volcanic ash is critical to recovery: buried seeds, stems, or roots, as well as burrowing rodents can sometimes survive when the ash is less than 1 m deep. Colonization of volcanic surfaces usually begins with the aerial dispersal of spiders and insects, followed often by nesting seabirds on island volcanoes (see Sections 3.11 and 4.2.1). These colonizers, plus wind-dispersed seeds or other plant parts capable of establishing, bring nutrients, trap sediments, and begin the process of re-vegetation and ecosystem recovery.

Recovery of organisms has been particularly well studied on volcanoes in Hawaiʻi, Iceland, New Zealand, and Japan. For instance, in Hawaiʻi, ecologists study changes across millennia using the whole island chain as a sequence of successively older surfaces. Studies of primary succession on Surtsey (Iceland) documented individual plants that have colonized since the island's formation in 1963 (see Chapter 4). Each new species of plant or animal is still noted during annual visits. Forests in the central North Island of New Zealand are still recovering from the Taupo eruption, which occurred 2000 years ago, and forests on Whakaari (White Island) never quite recover from on-going eruptions (Fig. 3.2). In Japan, one research emphasis has been on exploring the effect of microclimates on organisms on several recent volcanoes.

Fig. 3.2   Forests on Whakaari (White Island) in northern New Zealand are frequently damaged by toxic gases from the on-going eruption.

Volcanoes have been excellent places to study how communities and ecosystems assemble on barren surfaces and this knowledge can be applied to restoring analogous surfaces that humans create such as pavement, bulldozed areas, and mine wastes.

3.3 EARTHQUAKES

3.3.1 Characteristics and examples

Earthquakes affect all nine island groups and they normally occur along geological fault lines where sections of the hard, brittle surfaces of crustal plates grind past each other or collide and build up stress. When stress overcomes friction and the rocks suddenly slip, energy is released as an earthquake. This release of energy can occur during subduction of one plate under another (see Section 2.2). Alternatively, where plates are spreading apart, such as along the mid-Atlantic ridge where Iceland is located, earthquakes can be triggered by faults where there is lateral movement of two plates in opposite directions. Other causes of earthquakes include volcanoes and sudden movement of land or water, such as occurs during massive landslides or floods. Low-magnitude volcanic earthquakes are often associated with volcanic eruptions and movement of magma while higher-magnitude tectonic earthquakes occur in areas of structural weakness at the base of a volcano or deep in the crust. Human activities that can cause earthquakes include large construction projects, oil drilling, injection of fluids into wells, coal mining, and nuclear bomb testing. For example, many earthquakes were triggered when water first filled Lake Mead behind a newly built Hoover Dam in Nevada, USA. Earthquakes produce seismic waves that propagate through the earth from the point of origin and that have the power to deform rock. The strength of the wave dissipates with distance from the source. On the surface of the Earth, horizontal and vertical displacement can be several to many meters long. When this displacement results in either terrestrial landslides along coastlines or submarine landslides, tsunamis (tidal waves) can result.

The force of earthquakes is measured by several different scales. The Richter magnitude scale goes from 1 (a tremor so light that it is not usually noticed by humans) to 10 (a hypothetical value that has never been recorded, Table 3.2). Magnitude in this context is a measure of the power of the earthquake at its origin. The Richter scale, developed in 1935, is logarithmic, with a tenfold increase in magnitude with

Table 3.2. *Earthquake magnitude using the logarithmic Richter scale, global frequency of earthquakes at each interval, descriptions of typical damage, and examples from the last 100 years (dates in parentheses). Each increase in category represents a 10-fold increase in earthquake magnitude (energy produced at source of earthquake). BI = British Isles, CI = Canary Islands, HI = Hawai'i, IC = Iceland, JP = Japan, NZ = New Zealand, PR = Puerto Rico, TO = Tonga*

Richter magnitude score	Frequency (number per year)	Typical damage	Examples
1.0–1.9	2 920 000	Microearthquakes: not felt	All nine island groups
2.0–2.9	365 000	Minor: rarely felt	Most island groups
3.0–3.9	49 000	Minor: often felt, rarely causes damage	Most island groups
4.0–4.9	6 200	Light: shaking of loose indoor objects; minor damage	TO (2009); CI (2009); PR (2009)
5.0–5.9	800	Moderate: substantial local damage to poorly constructed buildings	HI (1975); CI (1982); BI (1984); TO (2009)
6.0–6.9	120	Strong: damage to 160 km distance	JM (1907); BI (1931); PR (1953); HI (2006); NZ (2007); IC (2008); JP (2008)
7.0–7.9	18	Major: serious damage to 200 km distance	IC (1910); NZ (1931); PR (1946); HI (1975); JP (2006); TO (2009)
8.0–8.9	1	Great: serious damage to 500 km distance	JM (1692); NZ (1855); PR (1946); HI (1975); JP (1994); TO (2009)
9.0–9.9	0.05 (1 every 20 years)	Great: serious damage to 5 000 km distance	Chile (1960); Alaska (1964); Indian Ocean (2004)
10+	Never recorded, rare	Epic: global scale destruction, global size fault line	

each increase in level. The Richter scale is still used (as in this book for comparisons among older eruptions – that are estimated prior to 1935) but is being replaced in the description of modern earthquakes by the moment magnitude scale that has similar numerical values to the Richter scale between 3 and 7. The moment magnitude scale uses a logarithmic scale with 30-fold increases. The damage caused at a given location is assessed with the Mercalli intensity scale that uses classes from 1 to 12 based on observations at each site. For example, people might notice an earthquake with a Mercalli value of 3 but not recognize it to be an earthquake; its vibrations feel similar to those from a passing truck. At a Mercalli value of 5, most people feel vibrations similar to a large train passing close by. At a Mercalli value of 7, it is difficult for people to stand and there is considerable structural damage, while at a Mercalli value of 9 and above, there is general panic and large buildings are destroyed. We now review the role of earthquakes in the nine island groups.

On the British Isles, Canary Islands, Jamaica, and Puerto Rico, moderate to strong earthquakes (>5.0 on the Richter scale) only occur on average once every 100 years. Notable earthquakes in the British Isles include one in 1580 in the Dover Straits that was between 5.3 and 5.9, and was associated with a destructive tsunami. Research on this earthquake was done in order to determine the appropriate design for the England-to-France tunnel. The highest magnitude (6.1) ever recorded near the British Isles occurred in the North Sea in 1931. Because strong earthquakes are so rare, citizens of the British Isles do not spend much time worrying about them. On the Canary Islands, where most volcanoes are extinct or dormant, damaging earthquakes are also uncommon. However, there is international concern that earthquakes at Cumbre Vieja, an active volcanic ridge on La Palma, could displace $500\,km^3$ of rock along a 2.5 km long fracture. If the mountain side collapsed in a massive landslide into the Atlantic Ocean, an enormous tsunami could devastate most Atlantic coastlines (including those of the British Isles, Iceland, Puerto Rico, and Jamaica), and could reach heights of tens of meters and travel many kilometers inland. However, there are also reports that this scenario represents a remote possibility; that future landslides will have only local effects; and that the slope will not collapse all at once.

The Caribbean islands of Jamaica and Puerto Rico are also not very often affected by earthquakes. However, strong earthquakes have occurred because there are several active fault zones in the Caribbean. About 16 000 years ago, for example, an earthquake triggered a

Fig. 3.3   Army magazine that partly subsided during the 1907
earthquake in Port Royal, Jamaica.

landslide in the Puerto Rico Trench that traversed at least 500 km and
had a volume of at least 100 km³. More recently, an 8.1 magnitude
earthquake near Puerto Rico (1946) produced a tsunami that killed
1600 people. In Jamaica, a strong earthquake in 1692 caused a sand
spit at the edge of the Kingston Harbor to collapse, killing several
thousand people and obliterating the pirates' capital of Port Royal.
Another earthquake in 1907 in the same area (Fig. 3.3) killed more
than 800 people. In 2010, nearby Haiti was struck by a 6.1 magnitude
earthquake that killed over 200 000 people and rendered over 1 million homeless.

Earthquakes are more frequent in Tonga (1 >5.0 magnitude every
10 years) and Iceland (1 >5.0 every year) than in the Caribbean Sea.
Two strong earthquakes (>7.0 magnitude) struck Tonga in March and
October 2009. Most of Iceland's earthquakes come from the friction
generated along geological faults or are associated with volcanic activity. In the last 2 years, earthquakes have briefly closed both a popular
thermal spa known as the Blue Lagoon and a geothermal plant. An
earthquake registering 6.3 occurred just 50 km offshore of Reykjavík in
2008, which generated some public concern but caused little damage.

The real hotspots for earthquake activity are Hawai'i (1.5 >5.0
magnitude every year), New Zealand (1.1 >5.0), and Japan (1–7 >5.0).
Thousands of earthquakes <5.0 occur annually in Hawai'i and most
are associated with the volcanic activity centered on and around the
Island of Hawai'i. Sometimes earthquakes are triggered by abrupt

Fig. 3.4   Landslides triggered by an earthquake in June 1994 and tree mortality in mountain beech forest, Avoca River Valley, South Island, New Zealand (see Box 3.1).

collapses along a coastal rift zone and associated with the accumulation of new lava. In 1975, two earthquakes on the southern coast of the Island of Hawai'i displaced the coastline 8 m horizontally and 3.5 m vertically and the summit of Kīlauea Volcano subsided 1.2 m. New Zealand and Japan are along the Pacific Ring of Fire where over ninety percent of the world's earthquakes occur. The Australians sometimes call New Zealand the "Shaky Islands" because they are so tectonically active (Fig. 3.4, Box 3.1). Since the 1840s, there have been about 140 earthquakes registering >5.0 on the Richter scale in New Zealand and eighteen >7.0. An 8.2 magnitude earthquake in 1855 displaced 5100 km² of surface 12 m horizontally on the site where the modern capital city of Wellington has since been built. Most recently, a 7.1 magnitude earthquake struck the city of Christchurch on the South Island in 2010, causing widespread damage to buildings and roads but no loss of life. Japan experiences about twenty percent of the world's earthquakes every year with an amazing six of the nineteen major earthquakes (>8.0) in the last 160 years. One in 1923 (8.3) occurred on the Kanto Plain where Tokyo is located and was very destructive, killing more than 100 000 people. In 1995 in Kobe, a 7.2 earthquake struck killing over 6000 people. Japanese citizens have clearly had to adapt as best they can to earthquakes as a common natural disaster.

Box 3.1. Walking on marbles

The Avoca River flows east from the Southern Alps on New Zealand's South Island, then south toward Lake Coleridge, and from there to the Rakaia River. The ranges through which the river and its side streams cut their course are steep, and the peaks are jagged. The forests that cover the lower slopes of the ranges are extraordinarily low in plant diversity; there is only one species of tree – mountain beech. The forests are stark and austere; there is very little in the understory. In 1999, Lawrence and I were part of a team measuring change in these forests. The forests are in a continuous process of recovery from past disturbances. Throughout the 1970s, the trees had been battered by strong winds, subject to branch break under particularly heavy snows, and attacked by bark beetles. Then, in June 1994, the head of the Avoca River was close to the epicenter of a 6.4 magnitude earthquake. By the standards of earthquakes in the Southern Alps, this was not exceptionally large, but the damage it caused was frightening enough. An eye witness account from near the epicenter describes the moment of the earthquake while he and his wife were walking in the valley. A tremendous roar, like the sound of a jet engine, resounded in the valley and the earth shook as they fell to the ground. He looked up to see boulders, the size of small cars, heaved out of the mountains above the treeline, and bouncing down the slopes, turning trees into showers of splinters each time they fell to Earth. One of their companions, in a small hut for hikers in the valley, described the wood-fired stove, after coming off its moorings, bouncing around inside the hut.

When Lawrence and I were in the valley five years later, close to the epicenter, the damage caused by the earthquake was obvious (Fig. 3.4). Landslides of broken rock created during the earthquake,

Box 3.1. (cont.)

up to 500 m across, extended from above tree line all the way to the valley floor. One of these landslides was large enough to dam a side stream and the debris had scoured the opposite side of the valley. The earthquake had happened in winter when the valley was snow-clad and, in spring after the snowmelt, the dam had burst, sending a wall of water down the side stream, tearing the forest off the stream banks. Walking up the side streams was unnerving. Log jams of trees that had fallen from the hillsides filled the valley floors. Raw walls of bedrock, exposed five years earlier, were scarred by debris. Sometimes the sound of rocks rupturing under strain could be heard. Slopes of shattered broken rock had to be negotiated, some frighteningly steep. Even within the standing forest, five years after the earthquake, the landscape was unstable. Broken rock littered the forest floor in many places, making walking a slippery business. Grabbing trees for handholds was risky because some were so poorly rooted that they would topple when pushed. Overall, about a quarter of the mountain beech trees close to the epicenter were killed during the earthquake and of the survivors about a fifth had been damaged by debris that had fallen. In a few places, groves of trees had been dislodged from where they had started life and slid downhill up to 200 m, and were growing well five years after the earthquake. There have been quite a few occasions since then, during work in valleys in the Southern Alps, I have thought to myself: "What would I do if an earthquake should happen now?". PJB

3.3.2 Ecological effects and responses

While some earthquakes directly disrupt organisms through displacement along the fault, much of the damage caused by earthquakes is indirect, via the tsunamis, floods, landslides, or fires that they can trigger. It is difficult to interpret the responses of plants and animals to earthquakes because of the combined direct and indirect effects of earthquakes and their unpredictable timing and severity. However, forest dynamics (such as in New Zealand and Japan) can be driven by earthquake effects that include direct shaking, soil movement, damage from boulders rolling down slope, and landslides. These types of earthquake effects killed about a quarter of the trees in a forest close to the epicenter of a 6.7 earthquake in New Zealand in 1994

(Box 3.1). Earthquakes that occur during movement of the large and active Alpine fault in New Zealand (Fig. 2.5), every 250 years on average, result in pulses of regeneration of many forest trees on landslides caused by the earthquakes in the mountains, or on the alluvial plains that are inundated with earthquake debris. The damage that earthquakes cause to homes, water, natural gas and sewage lines, oil tanks, and other infrastructures can indirectly impact the environment, which in turn affects human societies as they re-form after damaging earthquakes such as the one in 2010 in Port-au-Prince, Haiti.

3.4 EROSION

3.4.1 Characteristics and examples

Erosion, or the down slope movement of earth, is a natural consequence of geologic uplift, earthquakes, and destabilizing rainfall. Young mountain ranges are typically taller and more angular than older ones because they are not as eroded. Many of our island groups are mountainous and have various characteristics that promote slope erosion, including frequent earthquakes, high rainfall, and dense populations of humans that create road cuts (Fig. 3.5), log, and allow animals to overgraze slopes (Fig. 3.6). Each of these can trigger landslides, a common and highly visible form of erosion. Landslide intensity can be expressed in terms of depth of material removed, severity as the amount of organic matter or original soil profile removed, and extent as volume or area of displaced soil. Landslides often have discernible vertical zones including a slip face at the top where the landslide initiated, a chute where material passed through a relatively narrow zone in the middle, and a flatter and often wider deposition zone at the bottom. The deposition zone is usually the most fertile zone where colonization of plants and animals can be rapid. Only some landslides actually slide, and they can do that along a plane (planar slide) or by rotating (rotational slide). Landslides also fall (freefall of mass), topple (the mass tilts or rotates as a unit), flow (move with a fluid motion with or without substantial water), slump, or spread laterally.

In the British Isles, landslides are most common in the southern and central hills and along the coastline. Sometimes landslides occur where peat builds up over many centuries, as in Northern Ireland's Cuilcagh Mountain. Landslides are not common on the raised limestone islands of Tonga where most people live today but do occur on the newer volcanic islands. The Canary Islands have a long history of volcanic land

Fig. 3.5   Landslides are a common phenomenon along roadsides on tropical islands. This one is in eastern Puerto Rico and shows tree roots emerging from the cut bank.

Fig. 3.6   Soil erosion on slopes of Kaua'i, Hawai'i, which is likely due to overgrazing by goats.

Fig. 3.7   Landslide through montane rainforest in Garajonay National Park, La Gomera, Canary Islands.

building followed by landslides that eroded them (Fig. 3.7), sometimes as giant landslides involving hundreds of cubic kilometers of material. Three valleys on the northern slopes of Tenerife have been shaped by geologic uplift from the Las Canadas Volcano, followed by caldera collapses that have triggered large landslides. In Iceland, wind-driven soil erosion is common in the volcanic and glacial regions of the interior but landslides and the more prevalent (and deadly) snow avalanches predominate in the rugged hills and coastal lands dissected by fjords on the east and west coasts where the oldest rocks are found. Overgrazing is now a major cause of erosion in Iceland (see Section 5.9).

Hawai'i is frequently affected by landslides following heavy rainfall or earthquakes. On O'ahu, rainfalls of >76 mm in 6 hours usually trigger landslides, particularly along the mountain ridges. In the last 50 years, 1779 landslides have been recorded on O'ahu, with ones that cause significant damage occurring about once every 2 years. Other areas of frequent landslides in Hawai'i include the tall cliff faces on the north shore of Moloka'i, where an extensive landslide occurred in 1999, and the cliffs of northern Kaua'i. Landslides are also a very common feature of the mountains of Puerto Rico and Jamaica. High rainfall, steep slopes, old rocks with many fractures, and soil layers of varying permeability can promote landslides. Landslide-producing storms in Puerto Rico occur at an average rate of 1.2 per year and affect about 2–3% of forested landscapes in the Caribbean each century. This number can reach 7.5% in the mountainous regions. High rainfall events have triggered massive landsliding, including the record event of 865 mm of precipitation in a 10 hour period in 1979 in western Jamaica and a more "normal" 300 mm in 48 hours in the vicinity of Clarendon, Jamaica in 1986. An historically interesting landslide in Jamaica was the Judgement Cliff landslide in 1692, which followed a combination earthquake and cyclone. The name of the erosion scar left by the landslide was coined by Jamaican slaves because the land-slide killed a particularly cruel slave master; the slaves thought of the landslide as divine retribution for their mistreatment.

New Zealand's frequent landslides are due to the rapid uplift of the mountain ranges that produces high relief, steep slopes, and rocks weakened by folding and faulting. Earthquakes and abundant rain also influence landslide frequency. In general, New Zealanders worry about landslides when they receive >100 mm of rain in 24 hours. One notable landslide that occurred 13 000 years ago in the fjords on the west coast of the South Island displaced 27 km^3 of rock and soil over a 45 km^2 area. Another landslide in 1991 lowered the elevation of New Zealand's highest mountain, Aoraki (Mt. Cook), by 10 m.

Japan's numerous landslides predominate on unstable volcanic slopes, clay-rich slopes of highly weathered rocks, and hills of sedimentary rocks. However, most undisturbed, forested slopes do not slide, even when rainfall reaches >10 m per year, as on the island of Yakushima in the northern Ryukyu Islands. Nevertheless, landslides are so important to the highly concentrated Japanese population that local scientists and engineers have their own landslide society and journal. Most landslide research in Japan and elsewhere is focused on risk assessment and landslide geomorphology. One of the most extensive, and no doubt terrifying, of Japanese landslides occurred 6000 years ago in Matsushima Bay, Honshu where a mega-landslide covering approximately 1 million km^2 slid into the sea. As is common for large debris flows, this landslide left many hills and ridges. These topographical features now form the 200 islands distributed over the 150 km^2 bay that make it such a scenic place.

Plate 1   Lava from Kīlauea Volcano on the Island of Hawai'i has been flowing into the ocean for several decades, and has created several hundred hectares of new land.

Plate 2   Red cinder slope on Takachiho-dake Volcano (1574 m), Kirishima–Yaku National Park, southern Kyushu, Japan. Japan has one of the highest concentrations of active volcanoes in the world.

Plate 3 Vatnajökull National Park in southern Iceland is the largest national park in Europe, and encompasses Europe's largest glacier, Vatnajökull. In the background, Iceland's highest peak, Hvannadalshnúkur Volcano (2110 m) rises from the glacier.

Plate 4 Puffins on Latrabjarg Cliff in western Iceland. This 14 km-long cliff is several hundred meters tall and provides critical habitat for puffins and many other seabirds including fulmars, guillemots, and kittiwakes.

Plate 5 Forested northern slopes of the Blue Mountains, Jamaica. The Blue Mountains are Jamaica's highest mountains and are the main mountain range in the east of the island.

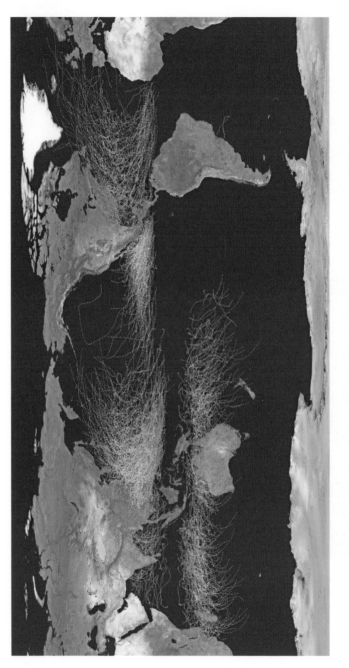

Plate 6   Tropical cyclone tracks. This map shows the tracks of all tropical cyclones that formed worldwide from 1985 to 2005. The points show the locations of the storms at 6-hourly intervals and use the following color scheme from the Saffir–Simpson Hurricane Scale: dark blue = tropical depression, light blue = tropical storm, light yellow = 1, dark yellow = 2, light orange = 3, dark orange = 4, red = 5. Accessed 15 July 2009 from: http://commons.wikimedia.org/wiki/File:Global_tropical_cyclone_tracks-edit2.jpg

Plate 7 *Wilkesia* or iliau, a close relative of the Hawaiian silverswords, is endemic to Kaua'i, Hawai'i. Most individuals of this rare plant are found in one 6 ha location where they are pollinated by both native and introduced bees.

Plate 8   A wētā in a forest remnant on Takapourewa (Stephens Island), Cook Strait, New Zealand. Wētās are flightless, nocturnal relatives of grasshoppers that are found in New Zealand and may have originally had similar ecological roles to rodents elsewhere, including the dispersal of native seeds.

Plate 9   Taro at Hermigua, La Gomera, Canary Islands. The wild progenitors of taro are found in Southeast Asia and had reached the Mediterranean Basin as a staple crop by the time of the Roman Empire. Whether its introduction to the Canary Islands dates from before the Spanish conquest is presently unknown.

Plate 10 Vava'u Group, Tonga showing human land uses including agriculture in the foreground and a village at center left.

Plate 11   Catch of valu and 'atu fish on Hunga, a small island in the Vava'u Group, Tonga. Subsistence fishing is still a part of Tongan culture, although sport fishing and commercial fishing for export are increasing.

Plate 12   A slate stone wall bordering an agricultural field on the coast of Cornwall, England. Stone walls are a ubiquitous feature of the British landscape and shelter many plant and animal species.

Plate 13   Sika deer grazing bamboo understory with no regeneration of the conifer overstory, Ohdaigahara, Mie, Japan. Sika deer populations in Japan have risen dramatically since the extinction of their native predators, especially wolves, in the early twentieth century.

Plate 14 Farming in Hermigua, La Gomera, Canary Islands. Stone terraces were built for agriculture on the steep hillsides of La Gomera, especially for bananas which were a boom and bust crop in the late nineteenth century. Many of these terrace gardens are abandoned.

Plate 15   Coffee fruits in a bowl, Blue Mountains, Jamaica. Coffee was a boom and bust crop in the Blue Mountains in the late eighteenth century. Currently, Blue Mountain coffee beans command some of the highest prices in the world.

Plate 16   Coal power station, Burton-on-Trent, England. The economy of the British Isles remains highly dependent on energy derived from burning coal. The UK now imports more coal, mostly from Australia and South Africa, than it exports.

3.4.2 Ecological effects and responses

Landslides have highly variable impacts on pre-existing organisms, from complete destruction or removal to simple displacement of intact "rafts" of soil and plants down slope. However, landslide scars are generally considered severe disturbances and recovery can therefore be slow. Where no soil or plant cover survives, landslide slopes are exposed to further erosion by wind or water. This typically occurs on the upper slip face or chute and recovery can be delayed for decades or even centuries. The lower deposition zone, that often contains a heterogeneous mixture of plant debris and fertile and infertile soils from above, can re-vegetate within several years, at least in favorably moist and warm tropical climates as found in Jamaica, Puerto Rico, Hawai'i, or Tonga. The increased light availability on a landslide compared to the low light availability under an undamaged forest canopy favors plants that invade quickly (or germinate from newly exposed buried seed pools). These pioneer plants can dominate and thereby delay forest succession if they grow densely enough or are superior competitors for light and nutrients. Efforts are now underway to link such ecological research with on-going restoration activities.

3.5 LAND BUILDING

3.5.1 Characteristics and examples

Wind or water-borne sediments that build (aggrade) land can be destructive to pre-existing ecosystems and when sediment deposition is rapid or abrupt, as during a flood, it causes a disturbance. In addition to tectonic earth building from volcanoes, sediments transported by water or wind can collect and build new land. There are at least four distinct ways in which this land building happens. First, water is the primary means of transport for coastal sediments. Islands are constantly changing in size as erosional processes along their coastlines are offset to varying degrees by aggradation. Beaches are an obvious place to examine erosion and sand build up. A single storm can remove the sand from an entire beach, only to move it along the shore and deposit it somewhere else. Large, coastal perturbations such as tsunamis (see Section 3.8) or storm surges during cyclones can move many cubic meters of sand in a single storm. Waves and tidal action also distribute sand. When beach sand is dry it is susceptible to wind transport and sand dunes can form inland from the beach. These dunes can grow as sand is added and can reach tens of meters in height. Sand is added to the gradually ascending windward side of a dune and then falls down the steep leeward side. Second, volcanic tephra can form sand that is redistributed by wind into craters or along coastal beaches. Third, primarily wind-dispersed sediments also come from inland dune fields such as the Sahara Desert and disperse to offshore islands such as the Canary Islands. Sediments from such sources can collect and build interior dunes with no coastal influence or water transport. Loess is a soil that comes from such aeolian sources and can reach depths of hundreds of meters. Finally, water and wind combine to transport sediment in a manner that links many types of disturbances. Tremendous quantities of sand or smaller particles such as silt or clay are transported to the coast in rivers and melt water from glaciers. Earthquakes trigger landslides that lead to increased river sediment loads; these earthquakes have been closely correlated with the formation of coastal dune ridges. The sediments that are washed down the rivers then augment shoreline sediments that in turn lead to more shoreline deposition and eventual increases in dune formation (Fig. 3.8). Dunes are not widespread but do occur on each of the island groups. We now discuss notable dunes from each group.

Tonga, Puerto Rico, Jamaica, and Hawai'i have wonderful, sandy beaches but dune formation is usually limited to within several

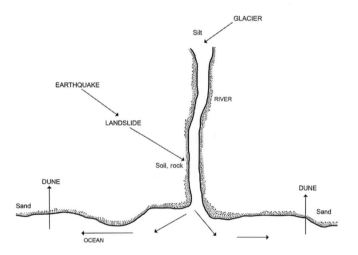

Fig. 3.8   Sediments from glaciers (silt) and earthquake-triggered landslides (soil and rock) flow downstream to the coast where they are further weathered and sorted, then deposited along shorelines as silt and sand that build dunes.

hundred meters of the shoreline. Tonga has relic dunes that were formed following historic earthquakes but they are now vegetated. The far northwestern corner of Puerto Rico has small dunes along a 40 km stretch of coastline. The dunes and accompanying sandy beaches are interspersed with rocky headlands and the beaches (the source of sand for the dunes) are frequently eroded by storms. Jamaica also has a few narrow dunes, but evidence of larger dunes, now stabilized, occur in areas such as the Palisadoes, which encloses most of Kingston Harbor. Hawai'i also has remnant dunes such as the Keopuolani Dunes near the Kahului Harbor on Maui and several black sand dunes behind black sand beaches of volcanic origin. One of the more extensive white sand dunes in Hawai'i is the 3.2 km-long Papohaku Dune on the west end of Moloka'i. The source of sand for this dune and a nearby inland dune system was probably the nearby Mo'omomi Sand Dune and now-inundated offshore sand fields.

The Canary Islands have the driest climate among the island groups and a plentiful source of sand from the nearby Sahara Desert; therefore there are more dunes than on the other small island groups. The largest are the Maspalomas Dunes on the southern coast of Las Palmas. These dunes extend 17 km along the shoreline and cover 4 km^2. Japan, despite its large land area, has few dunes. The most extensive are the Tottori Dunes near Tottori on the northern shore of Honshu. They cover 32 km^2 and are derived from sediments from the nearby Chugoku Mountains that are transported down the Sendai River to the ocean. The Tottori Dunes are tall, often reaching 40 m or more in height.

Iceland has numerous small, coastal sand dunes and several extensive ones. There are also interior dunes formed from erosion of sandy soils and dunes around volcanic areas including the colorful red and orange dunes in the Askya caldera in the north-central part of Iceland and black sand dunes near Reykjavík. Most Icelandic dunes, however, are fed by sediments coming down short rivers from melting glaciers. The most extensive of these dune fields is located on the 1000 km^2 Skeiðarársandur or sand plain on the south-central coast (Fig. 3.9). The dunes extend along most of the 40 km coastline of the sand plain and cover about 80 km^2. New Zealand's coast has extensive dunes along the west coasts of both the South and North Islands and a total of 1100 km of coasts with dunes (Fig. 3.10). Most of these are fairly narrow, but they extend 18 km inland and 200 km along the western coast of the North Island in Manawatu. Another large dune area is the 30-km long Farewell Spit in the far northwestern coast of the South

Fig. 3.9   A kettlehole formed from a melting ice chunk carried over 20 km out onto a huge floodplain (Skeiðarársandur) in southern Iceland by a glacial outburst flood; note glacier in the background.

Fig. 3.10   Dunes on the north side of Hokianga Harbor, North Island, New Zealand.

Island. Many New Zealand dunes are triggered by earthquakes that release material that is then washed ashore as dunes.

The British Isles have approximately 1600 km of coastal dunes scattered in all regions of the coastline; three-quarters of the total comes from dunes in Ireland, among which the 32 km² Magilligan Beach in

Northern Ireland is the largest. The tallest dunes in the British Isles are the 30-m tall Culbin Sands in eastern Scotland. The largest dune field in England is the 20-km long Ainsdale Dunes field along the Sefton Coast in northwestern England. It is aggrading but its source of sand is the nearby Formby Point that is eroding at 5 m per year. Braunton Burrows on the north coast of Devon in southwestern England is another interesting dune field because it is still very active, unlike many dunes in the British Isles that are stabilized by vegetation.

3.5.2 Ecological effects and responses

Coastal dunes erode as long as there is a supply of sand and wind to blow it inland. However, plants often colonize stabilized dunes within months of their formation, in part because the water table is often fairly high under dunes (water does not evaporate easily through such large pore spaces), and in part because of the lack of competition from other plants. Native plants, such as blue lyme grass from Iceland, are known as dune-builders. These plants collect mounds of sand around their bases and then grow up on to those mounds, thus stabilizing the dunes. Sea buckthorn is a common shrub on dunes in the British Isles and Japan that fixes nitrogen and is considered an important stabilizer. Marram grass is also native to Europe and planted widely for its ability to stabilize dunes. Other native plants and animals survive on the special habitat that dunes provide, including various spiders and beetles in New Zealand and lizards in the Canary Islands. Interdunal swales often have lagoons or ponds that migrating birds use for stopovers, such as the Canary Islands, when traveling between Europe and Africa. Humans have conflicting relationships with dunes. On the one hand, they want to stabilize them in order to avoid inundation of productive land from advancing dunes, while on the other hand, dunes are valued as habitats for distinct plants and animals. Dunes are also often seen as excellent beach-side locations for building houses despite their vulnerability to the disturbances that generate and maintain dunes.

3.6 TROPICAL CYCLONES

3.6.1 Characteristics and examples

Tropical cyclones are circular storms that generate over tropical oceans, between 5 and 30° latitude, where they obtain their energy from warm, moist, rising air (Plate 6). Tropical cyclones differ from temperate cyclones that are wedge- or oval-shaped storms formed between 40° and 60° latitude due to the convergence of warm (tropical) and cold (polar) air masses. In this discussion, cyclones will refer to tropical storms unless otherwise noted. A cyclone is characterized by rising, cooling air that loses its moisture as rain, and low atmospheric pressure. The trade winds and the spin of the Earth then help create these rotating storms. At the center of the spinning wind, the eye can have low barometric pressures, with average wind speeds of >118 km per hour. Because cyclones arise over oceans, islands are often the first to bear the brunt of their effects; islands are often affected more frequently than adjacent continents. Cyclones occur over all tropical oceans, but they are rare in the South Atlantic. Cyclone is a standard term used globally but regionally these storms have different names. They are called typhoons in the northwestern Pacific (Japan), cyclones in the southwestern Pacific (Tonga and New Zealand), and hurricanes in the north central Pacific (Hawai'i) and the Atlantic (Canary Islands, Puerto Rico, Jamaica). Once generated, they can move to latitudes as high as 45°. The intensity of cyclones is measured by the Saffir–Simpson Hurricane Wind Scale that relies on average sustained wind speed (speed maintained for at least 1 minute). This scale provides five categories of cyclones and two categories of less intense storms (Table 3.3).

The frequency of cyclones in the nine island groups varies from rare (Iceland and British Isles) to occasional (Canary Islands, Tonga, New Zealand, and Hawai'i), to frequent (Puerto Rico, Jamaica, and Japan), depending largely on the proximity of cyclone-generating oceans and trade winds. In the North Atlantic, cyclones are generated mostly off the west coast of Africa, move west with the trade winds, just miss South America and hit the Caribbean and eastern coasts of Central and North America. From there, some long-lasting cyclones move north and are pushed eastward by the westerly winds. We now discuss examples of cyclones from the island groups in increasing order of cyclone frequency.

The only cyclones that originate in the tropics that reach Iceland and the British Isles (Fig. 3.11) are those from the far western Atlantic that have degraded to extra-tropical storms or depressions by the time they reach Iceland or the British Isles. One example is Hurricane Ike (2008) that weakened to a tropical depression just west of Iceland in 2008. Nevertheless, this storm still caused 200 mm of rain in 24 hours in Reykjavík, 9-m high waves, and gusts up to

Table 3.3. *Saffir–Simpson cyclone scale of intensity based on wind speeds sustained for at least one minute. Each cyclone progresses through several wind speeds; examples represent wind speeds upon initial landfall at (or closest approach to) a particular island group that may not be the maximum wind speeds attained by a given cyclone. BI = British Isles, CI = Canary Islands, HI = Hawai'i, IC = Iceland, JP = Japan, NZ = New Zealand, PR = Puerto Rico, NZ = New Zealand*

Storm type	Wind speed (km/h)	Typical damage	Examples
Tropical depression	0–62	Possible minor wind damage to vegetation; localized flooding	Ike (IC) 2000
Tropical storm	63–118	Minor wind damage to vegetation, loose structures; more widespread flooding	Delta (CI) 2005; Jeanne (PR) 2004; Bola (NZ) 1988; Charley (BI) 1986
Category One	119–153	Moderate wind damage to vegetation, unanchored buildings, some uprooting of trees; extensive flooding	Hortense (PR) 1996; Luis (PR) 1995, Nina (1957) HI; Dot (HI) 1959; Iwa (HI) 1982
Category Two	154–177	Major wind damage to vegetation, windows, power lines; many uprooted trees; moderate to severe flooding	Georges (PR) 1998; Fernanda (HI) 1993
Category Three	178–209	Extensive wind damage to vegetation, buildings, power lines; extensive uprooting	Gilbert (JM) 1988; Uleki (HI) 1988; Waka (TO) 2001
Category Four	210–249	Extremely dangerous winds, massive damage to vegetation, buildings, power lines; nearly 100% uprooting	Hugo (PR) 1989; Iniki (HI) 1992; Tokage (JP) 2004
Category Five	>249	Catastrophic damage to vegetation, buildings; lengthy power outages likely	Ciriaco (PR) 1899; Giselle (NZ) 1968; Vera (JP) 1959

Fig. 3.11   Oak trees uprooted during a cyclone in October, 1987, near Farnborough, Kent, England.

144 km per hour. Similarly, Hurricane Charley (formerly a Category One cyclone when it reached North Carolina) passed over the British Isles in 1986 and caused damage to vegetation, flooding (setting a record rainfall in Ireland of 200 mm in 24 hours), and 11 deaths. The Canary Islands, close to the origin of the majority of Atlantic cyclones, receive only occasional tropical storms that approximate cyclones such as Tropical Storm Delta in 2005 that caused damage by uprooting trees, triggering landslides, and taking 19 lives in Tenerife and La Palma.

In the South Pacific, although New Zealand is only occasionally affected directly by cyclones, the results can be devastating. Cyclone Giselle came south from the Solomon Islands in 1968 with winds that ranged from 160 km per hour to a peak of 275 km per hour (Categories Two to Five). These were the strongest winds ever recorded in New Zealand. The cyclone and a storm coming up from Antarctica collided over Wellington and caused extensive damage; the inter-island ferry, the Wahine, sank with the loss of 51 passengers. As the cyclone proceeded through the South Island, plantation forests were destroyed, landslides closed roads, and rural villages were flooded.

Tonga has fewer cyclones than tropical islands further north, such as Samoa, but is affected by between one and two cyclones or tropical storms each year. In 2001, Cyclone Waka (Category Three) passed directly over Vava'u. Cyclones Keli (1997, Category Four) and Zoe (2002, Category Five) caused extensive damage to Tonga, although they did not directly hit Tonga. Hawai'i is at the far west end of the path of cyclones generated in the eastern Pacific off the coasts of Central American and southern Mexico, so cyclones often dissipate before reaching it. Hawai'i has been directly impacted by only two major cyclones in the last 50 years, Hurricanes Dot (Category One) in 1959 and Iniki (Category Four) in 1992. The greatest impact occurred on Kaua'i in each case.

Cyclones are a common occurrence in the Caribbean Sea. The word hurricane comes from Hanaka, the Taíno's storm god, who was thought to reside in the Luquillo Mountains of Puerto Rico and protect the Taíno from cyclones. Indeed, the mountains sometimes do cause cyclones to lose power or swerve. On average, five cyclones impact the Caribbean Sea each year. During the last 157 years (1852–2009), Puerto Rico has been directly hit by 17 cyclones or one every 9.2 years on average. Jamaica has been hit by 14, or one every 11.2 years on average (Fig. 3.12). Historical records of cyclones in the Caribbean Sea start in 1502 when Christopher Columbus warned the Spanish governor of Hispaniola to no avail of a big storm that ended up sinking 21 Spanish ships in Mona Passage between Hispaniola and Puerto Rico. The most recent hurricanes that have caused major ecological damage were Hurricane Gilbert (Category Three, Jamaica, 1988) and Hurricane Hugo (Category Four, Puerto Rico, 1989). Hurricane Ivan (Category Five, 2004, Box 3.2) just missed Jamaica but still buffeted it with cyclone-force winds.

Fig. 3.12   Airplane damage at Kingston, Jamaica airport caused by Hurricane Gilbert, 1988.

Box 3.2. Hurricane Ivan

The hurricane season in the Caribbean was quite a busy one in 2004, when I was working in the forests in the Blue Mountains of Jamaica. Forecasting hurricane tracks is never certain but it has improved in quality markedly over the last decades. The news came that a powerful hurricane was heading toward Jamaica while I was still in the mountains, so my colleagues and I decided that the wise course of action was to pack up gear from the remote field station and head to the relative safety of a friend's house in Kingston. With a day to prepare before the hurricane's expected landfall, I joined queues at supermarkets for food, and helped put up shutters and tape across glass windowpanes. I spent the late afternoon cutting low branches of trees close to the house to ensure that they caused no damage. As evening approached, storm clouds darkened the sky, and the radio and television reports confirmed that the track of the hurricane, which was gathering in strength as it moved, was on a direct course toward Kingston. Warnings to evacuate were given to those who lived near the coast or in low-lying areas. There was nothing to do but wait and hope for the best.

  After dark, the wind began to grow in strength and the rain, which had begun as night fell, became heavier. It was a long, stormy night. Dawn broke grey and wet. The strong winds, which had blown strongly all night, began to lessen. I woke, with the rest of Kingston, to no electricity, because all of the power lines were down. Water had seeped into the house through gaps in the windows or doors. Shredded pieces of leaves covered all the surfaces outside (Fig. 3.14). Broken tree limbs littered the yard, and fallen trees were evident in neighboring yards. But the eye of the hurricane had missed Kingston. During the night, the track of Hurricane Ivan had swerved abruptly and, while only 50 km away from making landfall, headed westward along the southern coast of Jamaica. The city had been spared the worst of this Category Five hurricane. Within days the folklore had already begun: the city of Kingston was too frightening and had scared the hurricane off! Yet even being side-swiped by this hurricane, the sixth most intense on record, was bad enough for the island. I walked along the streets of the neighborhood, negotiating fallen trees and power poles, to look into the ravine of the nearby Hope

Box 3.2. (cont.)

River Valley. The valley had been scoured away by the intense rainfall and many homes had been swept away. The roads into the mountains had been undermined by swollen rivers or covered in the debris of landslides, so I was not going back to work any time soon. At least 18 000 people were left homeless and 17 people had been killed by flash floods or the storm surge. PJB

Cyclones in the northwestern Pacific are the largest and most intense in the world. They originate in the Central Pacific and move west toward the Philippines, then generally veer northward through Taiwan to the Chinese coast and to Japan and Korea. One-third of all cyclones are formed in the northwestern Pacific each year, including 26 of the 30 cyclones with the lowest atmospheric pressure. Japan gets hit each year by 3 to 10 of the approximately 28 cyclones that are formed in that region. The Ryukyu Islands in far southern Japan (including Okinawa) get hit by the bulk of these cyclones, but all Japanese islands have been struck. Most cyclones lose their strength as they move northward, turning into tropical storms by the time they reach Honshu and Hokkaido. Yet cyclones are frequent enough that many large Japanese cities are built on the leeward (northern) side of bays and islands because cyclones usually arrive from the south. The most famous historical cyclones in Japan were the Kamikaze (= "divine wind") cyclones of 1281 (Box 3.3). The most recent of nine Category Five cyclones to directly impact Japan in the last 50 years was Typhoon Yagi (2006); the most intense was Typhoon Tip (1979; lowest atmospheric pressure ever recorded: 870 hectopascals); the one with the highest wind speed was Typhoon Vera (1959; 315 km per hour); and the most destructive to humans was Isewan (1959; 5000 people killed). In recent decades, deaths have been minimized due to better building construction and more advance warning.

Box 3.3. Kamikaze cyclones repel Mongol invaders

In both 1274 and 1281, Mongols attempted to invade Japan but both times they were thwarted in part by cyclones. After the Mongols completed the occupation of Korea in 1259 and after the ascension of Kublai to leadership as the Great Khan in 1260, several groups of emissaries were sent to Japan, demanding its surrender.

Box 3.3. (cont.)

The Japanese refused and began to prepare for an invasion. In 1274, a Chinese navy of at least 800 boats and 15 000 men took off from Korea and landed in Hakata Bay in Kyushu. After initially successful advances against the outnumbered and more poorly equipped Japanese army, a cyclone arrived and the Chinese forces retreated to their ships. However, the storm destroyed several hundred ships and the Japanese, who excelled in single combat, harried the Chinese from their smaller, more agile boats until the Chinese navy retreated. Again, after further Chinese emissaries demanding surrender were beheaded by the Japanese, another large navy was assembled and a second attempted invasion occurred in 1281. However, the Japanese had built many protective fortifications along their coasts and were better able to repel the invaders. Retreating once again to their ships, the huge Chinese navy lingered for several months off the Japanese coast, increasingly demoralized and with deteriorating supplies. Once again, a cyclone arrived and caused most of the hastily and poorly made Chinese ships to capsize. The Mongols never attempted another invasion and the cyclones were henceforth considered "divine winds" or kamikaze. The cyclones that batter Japan every year played a major role in limiting Mongol expansion and reinforcing a sense of invincibility among the Japanese. LRW

3.6.2 Ecological effects and responses

Cyclones have short- and long-term effects on landscapes and the biota. Immediate effects on the landscape depend on wind speed, size of the storm, and path and residency time across the landscape. Local factors that sometimes ameliorate cyclone impacts include mountain ranges and leeward slopes. Floods and landslides are often triggered by the heavy rainfall that accompanies most cyclones. For example, Tokage in Japan (2004) triggered more than 280 landslides. Cyclone-induced landslides are also common in Jamaica and Puerto Rico. Damage to the vegetation includes defoliation, branch loss, and snapping or uprooting of tree trunks (Figs. 3.13 and 3.14). An analysis of 85 cyclones that struck Puerto Rico since 1508 concluded that there was about a 5-year return interval for loss of leaves and branches and limited felling of trees, and a 15–150-year return interval for extensive areas of felled trees. Animals either try to escape or hide from cyclones. In the first decade following one well-studied cyclone in Puerto Rico (Hurricane

Fig. 3.13   Coastal forest damaged by Cyclone Waka in 2001 near Holonga on the north coast of Vava'u, Tonga. This photo, taken six months after the hurricane, shows some plants growing back into the opening created by the hurricane.

Fig. 3.14   Damaged trees the morning after Hurricane Ivan, 2004, in Mona, Kingston, Jamaica (see Box 3.2).

Hugo, 1989), stick insects, fruit-eating birds, and juvenile frogs declined while other insects, spiders, and insect-eating birds increased. Adult frogs and lizard species showed no overall effects. Sometimes birds are blown long distances, as when North American birds appeared in the British Isles following Hurricanes Frances (2004) and Gustav (2008).

Longer-term effects include gradual increases in tree mortality in addition to those not killed immediately. Gaps in the forest canopy or other vegetation allow new colonists to compete with established vegetation as plant recovery begins. Complete recovery of pre-cyclone plant communities takes at least 50–150 years on islands in warm, wet climates favorable to plant growth, but can take several centuries in dry or cool climates. Animals that survive the disturbance are also presented with opportunities to colonize newly denuded landscapes if the species excel in rapid dispersal and reproduction. However, adjustments to an altered environment are likely. For example, lizards living in upper tree branches may have to adapt to sharing the forest floor with other species until the vegetation grows back. Similarly, branched corals take at least a decade to develop, but cyclone intervals are shorter than that, so that the corals are typically recovering from the latest cyclone. Unfortunately, the future of corals in the Caribbean (and elsewhere) is bleak because they face not only cyclones but sea-level rise, ocean acidification, ocean warming, bleaching, dynamiting, and harvesting.

3.7 FLOODS

3.7.1 Characteristics and examples

Floods can damage plants and animals through both physical destruction and habitat alteration such as by erosion or burial under sediments. Floods are a common disturbance on most of the nine island groups. Humans are particularly vulnerable to floods because people favor riverbanks, floodplains, and coastal plains for living and for agriculture. People are not only attracted to the fertile soils that result from floods but also use rivers for transportation, fishing, and recreation. Floods occur mostly in rivers or along coasts but sheet flooding can occur on slopes such as the eastern slopes of Mauna Kea on Hawai'i that receive more than 7 m of rain each year. Floods also occur when glaciers melt rapidly, as from a volcanic eruption; when dams break that are made from either natural debris or constructed by humans; or when lakes, including those in volcanic craters, overflow. We discuss each of these types of floods as they apply to the nine island groups before summarizing their effects and the response of the biota.

Glacial outburst floods

A very dramatic type of flood is a glacial outburst flood or jökulhlaup. These floods can have enormous water volume and there is evidence of their magnitude and terrain-sculpting abilities from geological records, particularly during the melting of continental ice sheets at the end of ice ages. Glacial outbursts occur whenever a subglacial lake overflows or breaks out in a sudden release of massive amounts of water. The best known are the outbursts caused by eruptions of volcanoes under Iceland's glaciers. Subglacial lakes form from the heat of the eruption, and then eventually break out by melting a tunnel through the ice and emerging to spread out over wide floodplains. Icebergs can be transported in these huge floods that can last between one day and several weeks or months. As the icebergs melt, depressions in the terrain are formed (kettleholes, Fig. 3.9). Peak flows in Iceland have ranged from 50 000 m³ per second (1996) to 300 000 m³ per second in a flood that occurred in 1918 to an estimated 500 000 m³ per second in a flood that occurred 2500 years ago. New Zealand is the only other one of the island groups with glaciers (Fig. 3.15). Outbursts occur in New Zealand, such as at the Fox Glacier, which has been melting for the past 60 years; water trapped in tunnels below the Fox Glacier occasionally bursts out when its icy cap melts.

Fig. 3.15   Maud Glacier and lakes at the head of the Godley River, Mount Cook National Park, South Island, New Zealand. This glacier is on the eastern (leeward) side of the Southern Alps, where many glaciers are retreating.

Other outbursts

Debris dams can temporarily stop river flow but cause floods when they break. Landslide-caused dams are common in rugged mountain ranges such as those found in New Zealand and Japan. Dams also occur from sediments, logs, or other material transported during previous floods. Lakes can be drained by earthquakes that create new outlets or by landslides or ice chunks that fall into a lake, pushing water over the edge. Lakes can also simply fill up and then break through their confining walls, particularly when those walls are not very stable. Volcanic crater lakes can breach their rims with catastrophic consequences, as has occurred in New Zealand. When crater lakes in the central North Island have breached rims of ash, for example, enormous floods called lahars have occurred. These floods can dump 20–60 km³ of water into drainages and deposit ash and large rocks many kilometers downstream. When a lake in the crater of Ruapehu breached the crater rim in 1953, a lahar destroyed a railway bridge minutes before an express train was due to cross; 151 people were killed in the ensuing derailment. Dramatic floods like these were a big part of shaping current landmasses and drainages on many islands. For instance, any earlier land connections between the British Isles and mainland Europe were probably severed in massive floods that carved the English Channel 450 000 and 200 000 years ago.

Coastal flooding

Islands are particularly vulnerable to coastal flooding because of their proportionally long coastlines relative to land area and their susceptibility to storms from any direction. Coastal flooding can occur from storm surge, when cyclones or other high winds push the waves onto the shore. Another cause comes from a combination of high tide and high river levels. One notable storm surge hit the east coast of the British Isles in 1953. Called the Big Flood, it was the worst natural disaster to occur in the British Isles during the twentieth century. The flood was spawned by a temperate cyclone and funneled down the shallow North Sea with winds gusting to 180 km per hour. The low atmospheric pressure associated with the storm helped lift the North Sea 2 m. The storm surge went several hundred meters inland in low-lying places, bringing beach sediments and toppling vegetation. Poorly constructed, temporary buildings were demolished and 307 people lost their lives. Tsunamis are usually associated with coastal flooding and are discussed in Section 3.8.

River flooding

Rivers flood when there is a period of heavy or prolonged precipitation somewhere in the river catchment, when accumulated snow or ice melts, or when a dam or flood control wall is breached. Saturated soils that no longer absorb water also promote flooding. A flash flood occurs when a river rises quickly (Fig. 3.16). Arid regions such as the Canary Islands are particularly vulnerable to flash floods (such as occurred in November 2001 and March 2002) because the dry soils do not easily absorb precipitation and runoff is rapid.

Steep, short drainages that empty onto coastal floodplains are typical of most of the nine island groups. The mountainous terrain receives more precipitation than the lowlands, but down slope transfer of water and sediments can be very rapid and sometimes unexpected. For example, peak stream flow can occur less than 1 hour after peak rainfall in Hawai'i, greatly reducing the time available to evacuate flood zones.

The intensity of rainfall is a key factor in determining if a flood will occur. Jamaica has a weak, bimodal pattern in annual precipitation

Fig. 3.16   Flood stage of the Whanganui River on the West Coast of the South Island of New Zealand. Within hours the track along this river was covered meters deep in rushing water. Most valleys in this region receive between 6 and 11.5 m of rain each year, and much of this falls in storms that last one to three days, like the one here.

Table 3.4. *Notable storms with high rainfall intensity that caused extensive flooding. For other notable storms see Section 3.4 on landslides*

Location	Year	Rainfall (mm)	Time period (hours)
Roxburgh, New Zealand	1992	80	0.75
Tokai, Japan	2000	97	1
Southeastern Puerto Rico	2008	102	1
Okazaki, Japan	2008	150	1
Somerset, British Isles	1968	125	1.5
Boscastle, British Isles	2004	200	4
Western Jamaica	1986	865	10
Invercargill, New Zealand	1984	85	24
Ireland, British Isles[a]	1986	200	24
Reykjavík, Iceland[b]	2008	200	24
Somerset, British Isles	1924	238	24
Bruton, British Isles	1917	243	24
Tongatapu, Tonga	2000	289	24
Okazaki, Japan	2008	302	24
Gisborne, New Zealand[c]	1988	514	24
Nagoya, Japan	2000	534	24
Hilo, Hawai'i	2000	566	24
Puerto Rico[d]	1975	584	24
Cerro Maravilla, Puerto Rico	1985	624	24
Hilo, Hawai'i	2000	665	24
Lower North Island, New Zealand	2004	300	48
Luquillo Mountains, Puerto Rico[e]	1998	>700	48
Vega Baja, Puerto Rico	2001	508	72
Santa Isabel, Puerto Rico	2008	760	72
Gisborne, New Zealand	1988	>900	72

Notes: [a] Hurricane Charley remnant; [b] Hurricane Ike remnant; [c] Cyclone Bola; [d] Hurricane Eloise; [e] Hurricane Georges.

(peaks in October and May) but a unimodal peak for flooding (half of all floods occur in October). Therefore, it is not the magnitude of monthly rainfall that determines flooding but its distribution during the month. Rainfall intensity (millimeters per hour) usually correlates with flood intensity (height of water or flow in cubic meters per second), although other factors (notably prior degree of soil saturation) are also important. Some notable storm intensities are listed in Table 3.4, with the most impressive record coming from western Jamaica in 1986 when 865 mm fell in just 10 hours. Hawai'i's storms are also strongly seasonal (mostly during the wet season from October to April) and Hawai'i has about eight flash floods each year (Box 3.4). Sometimes meteorological conditions simply contribute to an unusually wet year, as in the British Isles in 2000 when precipitation 200% above normal led to widespread flooding.

Box 3.4. Tropical rainstorms are frequent but short

The average storm in the Caribbean National Forest, Puerto Rico lasts only 20 minutes. Just sit under a wide, leafy tree and wait. However, storms can settle over an island and produce copious rain. While living in Hawai'i Volcanoes National Park, I experienced a storm that dumped 230 mm of rain in 24 hours and continued for several weeks, totaling 760 mm in 30 days. Puddles are not common on porous lava, but this storm produced short-lived puddles to splash in as well as flash floods in the lowlands. LRW

3.7.2 Ecological effects and responses

Floods transport vast amounts of sediment downstream (up to nearly 1 km³ as in the 1918 Iceland glacial outburst flood), thereby increasing soil fertility in floodplains and nearby coastal waters. Floods also rearrange existing landscapes by scouring new channels, forming new banks, and straightening curves. Considering that 1 m³ of water weighs 1000 kg, it is apparent what kind of damage the momentum of a flood can cause. The addition of sediments makes a flood into a debris flow that is even more destructive – akin to wet concrete. With this power, huge boulders are moved considerable distances and bridges and roads are often damaged or destroyed. The response of the natural world (and humans) to flooding is to reinvade flooded areas. Fast-moving and fast-growing colonists begin the process of succession on new deposits. Fine silts tend to be colonized by grasses while rockier floodplain surfaces often are colonized by shrubs or trees. Most early plant colonists can withstand minor floods but will die during 20 or 100 year events. If major floods do not return for a century or more, less flood-tolerant plants and animals will eventually establish. Of course, any organism living on a floodplain, an area that is usually wider than the river channel, is vulnerable to the next flood.

3.8 TSUNAMIS

3.8.1 Characteristics and examples

A tsunami, which means harbor wave in Japanese, is a natural disturbance that potentially impacts any coastline. Each of the island groups except Iceland and the British Isles is frequently impacted by tsunamis. The coastlines of Japan, New Zealand, and Hawai'i are damaged by 30% of the approximately 20 damaging tsunamis that occur each decade. Tsunamis are triggered by forces that cause a sudden change in sea level. Ninety percent of tsunamis are caused by submarine earthquakes and less than 1% of submarine earthquakes trigger tsunamis. Other causes of tsunamis include comets and meteorites, volcanoes, landslides, avalanches, glacial outbursts (jökulhlaups), sudden violent rains, and the bursting of dams. Tsunamis have less than 10% of the energy of the earthquake. When the earthquake is less than 8.0 magnitude the impact of the tsunami is limited to several hundred kilometers from the epicenter and waves are generated that are less than 2 m tall. A 1993 earthquake (7.8) in the Sea of Japan, for example, mostly impacted the western coasts of Japan up to 200 km from the source. In 2009, an earthquake (8.0) off the coast of Samoa devastated Samoa with four large waves and also killed nine people on the northern Tongan island of Niuatoputapu. In contrast, higher-magnitude, submarine earthquakes generate tsunamis that can impact islands anywhere in the same ocean. For example, the strongest earthquake ever recorded (off the coast of northern Chile in 1960) reached 9.5 magnitude and impacted islands across the Pacific Ocean, including Hilo, Hawai'i (10 000 km distance reached 15 hours later, 11 m tall wave, 67 people killed) and Honshu, Japan (18 000 km in 22 hours, 6 m wave, 190 people killed).

The shape of a coastline determines the impact of a tsunami, in addition to the direction and distance from the trigger event. Islands that have gradually sloping harbor basins, such as Hilo, Hawai'i are especially vulnerable because the wave can build to dangerous levels as it funnels in and up the harbor floor. The same tsunami would hardly be noticeable in the middle of the ocean because the wave length (distance between wave peaks) is much greater (up to 10 km or more) and the peaks much shorter (several centimeters or decimeters). Volcanic eruptions and earthquakes throughout the whole Pacific Ring of Fire can affect Hilo. Earthquakes in 1868, 1946, and 1960 produced destructive waves in Hilo ranging from 8–15 m tall. Not all tsunamis are so tall, however. In 2010, an 8.8 moment magnitude earthquake in Chile produced maximum wave heights of <1 m in Hilo (Box 3.5).

Box 3.5. The Pacific Ocean is not that big

I found that the Pacific Ocean was not that big when a tsunami practically chased me from Chile to Hawai'i. Leaving southern Chile 28 hours before an 8.8 moment magnitude earthquake hit, I was in Hawai'i for only 19 hours before the tsunami generated by that earthquake reached Honolulu, Hawai'i. Unlike the unfortunate Chileans who experienced a tsunami wave reaching several meters that compounded the disaster caused by the earthquake, Hawai'i received waves that peaked at about 1 m on Maui. Expectations were for waves to reach at least several meters so thousands of residents moved to the hills, tourists were vertically evacuated to higher floors in the coastal hotels, and busy Honolulu turned into a ghost town. I watched from a seaside perch (on a ninth floor) as three surges of <50 cm – easily within the range of normal tidal fluctuations – came and then receded in the space of about one hour. Few complained about the disruption of their day, however, as Hawaiian locals remember the loss of life and property that occurred in Hilo during tsunamis in 1946 and 1960. LRW

Both geological evidence (such as huge limestone rocks perched tens of meters above sea cliffs) and cultural legends from Aboriginals suggest that a comet struck the Tasman Sea southwest of New Zealand about 8000 years ago and produced a large tsunami in the region. Many other lines of tentative evidence support this suggestion, including evidence of clusters of skeletons of moa in swamps (as though these flightless birds had fled en masse) and carbon dates of extensive fires in uninhabited uplands that might have resulted from widespread firestorms.

Five percent of all tsunamis are triggered by volcanoes. The best documented volcanic tsunami formed from the eruption of Krakatau Volcano in Indonesia in 1883. The Sakurajima Volcano in Japan (Fig. 3.17) erupted in 1780 and its 6-m high wave was generated by an underwater explosion. Other Japanese tsunamis were created by debris avalanches formed from collapsing volcanic slopes. For example, the 1792 eruption of Unzen Volcano in Kyushu, Japan sloughed 0.34 km^3 of material into the ocean and generated a tsunami up to 55-m tall. There have been five other volcanic tsunamis in Japan since the early 1640s. In 1693, Iceland's Mt. Hekla created a

Fig. 3.17   Sakurajima Volcano, near Kagoshima, Kyushu, Japan. This
volcano has had major lava eruptions in 1476, 1779, 1914, and 1946.
Ash from this volcano often falls over the city of Kagoshima.

tsunami from a volcanic earthquake and in the Caribbean, St. Vincent
and Pelée both generated tsunamis in 1902 from pyroclastic flows.
Tsunamis are not always limited to oceans, as Krakatau's blast in 1883
generated a pressure wave that caused a 0.5 m-tall tsunami on Lake
Taupo in New Zealand, 4700 km away! An eruption of Ruapehu in
New Zealand in the 1980s generated tsunami waves that were about
2-m tall on Lake Taupo.

Some of the world's largest tsunamis are caused by massive
submarine landslides. Two landslides occurred 8000 years ago in the
North Sea between Norway and Iceland. An estimated 5580 km^3 of
sediment collapsed 2 km into the depths of the North Sea across an
area about the size of Iceland (110 000 km^2), causing the Storegga tsunamis that had waves up to 30-m tall along the coasts of the British
Isles and Iceland. Evidence for this event comes from sand containing diatoms that was washed ashore to those heights. Large submarine debris flows that reached 100 km in length and are up to 2 km
thick with volumes of 400–5000 km^3 occurred several hundred thousand years ago around several oceanic islands including the Canary
Islands and Hawai'i. These submarine landslides were triggered by
the instability caused by the growing volcanoes. The tsunami generated by the Canary Island landslide likely deposited ridges on the
Bahamas, and the Hawaiian tsunami may be responsible for the
365-m tall wave on nearby Lāna'i. Even relatively smaller underwater
landslides can have important localized impacts. Large, house-sized

rocks called the Coral Boulders located near the village of Fahefa on Tongatapu, Tonga are up to 20 m above current sea level and 400 m inland. Models suggest that they were placed there by a 40-m tall tsunami generated from a nearby 1-km^3 submarine landslide that occurred about 120 000 years ago.

Terrestrial landslides, especially large ones triggered by volcanoes or earthquakes, can also cause tsunamis when the material ends up in the ocean (see Section 3.5). The tallest known tsunami occurred when an 8.3 magnitude earthquake triggered a landslide in southeastern Alaska in 1958 sending a wave with a height of 524 m onto the opposite side of Lituya Bay. Glacial outburst floods can also cause tsunamis (see Section 3.7.1). A large flood of this kind in Iceland in 1918 produced a 1–5-m tall tsunami that reached the offshore Westmann Islands.

3.8.2 Ecological effects and responses

Tsunamis have multiple effects on the structure of coastal landscapes, including several types of depositional impacts such as layers of sand, mud, and diatoms. Tsunamis also have erosional impacts, including sculpting cliffs, caves, arches, and rock towers called fluted stacks as well as forming lagoons and shaping coastal hills. Effects on the biota and responses to tsunamis are poorly documented. The initial damage is similar to the combined effects of a saltwater flood or a cyclone that breaks up coral reefs, destroys mangrove forests, buries coastal vegetation (including crops) in salty sediments, and defoliates trees. Intact mangrove forests and coral reefs provide some protection from tsunami damage. Marine and terrestrial animals are also clearly affected, with highest mortality for immobile species such as corals, shellfish, burrowing animals, and soil organisms. The terrestrial impacts can occur for several to tens of kilometers inland. Other effects include the loss of human lives and fires from broken gas pipelines, leaking fuel tanks, and electrical short circuits. The destruction varies with the height and strength of the waves, the behaviors of people and animals, and the timing of the life cycles of plants.

Responses of organisms (other than humans) to tsunamis have rarely been studied. The recent tsunami in the Indian Ocean in 2004 heavily damaged coral reefs along the coast. Within just several years, rapid growth of young coral has been observed. Perhaps most importantly, some residents are now minimizing destructive fishing techniques and even assisting in transplanting coral to damaged areas. On land, tropical climates support rapid regrowth of vegetation and recovery of plants such as palm trees that often survive tsunamis. Where homes are abandoned, they can quickly be enveloped in vegetation. In climates less favorable to growth than tropical climates, recovery is naturally slower. Of course, replacement of old, slow-growing trees or any vegetation growth on severely salt-damaged soils can take centuries. The most permanent signatures of a tsunami are the physical changes to the landscape.

3.9 DROUGHTS

3.9.1 Characteristics and examples

Droughts can be considered disturbances when they are relatively sudden and unexpected. Droughts, unlike floods, hurricanes, and other disturbances, have no clear beginning. They are only recognizable after a period of time. Droughts can be defined in meteorological terms as lower than expected rainfall or higher than expected evaporation; in hydrological terms as reduction in stream flow; in soil water or agricultural terms as damaging to crops; or even in sociological terms as a loss of jobs related to an affected industry, such as agriculture. Meteorological causes of droughts include variations in wind patterns and sea temperatures such as the effects of El Niño weather that brings drought to Tonga and Hawai'i. Droughts that are seasonal or annual would not commonly be considered a disturbance. Droughts are often associated with dry climates and are defined relative to the normal precipitation. Several weeks without rainfall in wet parts of Puerto Rico, for example, is a drought. The nine island groups have oceanic climates with high humidity and generally ample rainfall. Nonetheless, each of the island groups does experience drought (presented below in approximate order of increasing susceptibility to drought).

   Iceland is perhaps least affected by droughts, not only because of frequent rainfall but because of cool air temperatures that reduce evaporation. The British Isles also have relatively reliable rainfall. Analyses of droughts in the British Isles are facilitated by a very thorough system of rain gauges (over 8000 in all) and one of the longest records of precipitation, going back to 1677. Droughts from periods of 10 days without rain in the British Isles occur on average every year, from 20 days without rain every 10 years, and from 30 or more days without rain every 25–30 years.

Hot, wet climates such as found on parts of Puerto Rico, Jamaica, Tonga, and Hawai'i are not highly susceptible to droughts but their large variations in precipitation and topography (all but Tonga have a pronounced leeward or dry side) make the drier sides particularly susceptible to drought. Puerto Rico and Jamaica have had severe droughts about once every 25 years. Tonga's southern islands are most susceptible, particularly during El Niño years such as 1997/1998. Hawai'i varies the most in rainfall (see Table 2.1) and has had 17 major droughts in the last century. Every part of the island group has been affected. The 1983–1985 and 2009–2010 droughts were El Niño-related.

The largely temperate climates of New Zealand and Japan are not particularly drought-sensitive but do experience droughts about once every decade. In New Zealand, an absolute drought is defined as 15 days without rain, while a partial drought is defined as 30 days with less than 2 mm of rain per day, on average. The Canary Islands are the driest and most drought-susceptible of the nine island groups. Data from pine tree growth rings suggest that there have been about 20 major droughts in the Canary Islands in the last century.

3.9.2 Ecological effects and responses

Drought adaptations by plants and animals include a reduction of metabolic activity, internal storage of water, or use of a permanent water source (such as a spring for an animal or groundwater for a plant). Several weeks without rain in a New Zealand grassland is not uncommon, and although the aboveground leaves dry out, the plant can resprout after rains resume. Plant distributions on the Canary Islands are partly determined by their relative drought tolerance but also by their tolerance to cold at upper elevations. Droughts that lead to forest or brush fires (as in Hawai'i in 1901 and 1905) favor plant species that are fire-adapted and many of these are introduced rather than native (see Section 6.6.2). Current emphases on drought effects focus around human needs for drinking and for agriculture. Water-intensive crops such as sugar cane do not survive droughts. There are both ecological and social consequences when streams dry out because streams normally fill reservoirs, irrigate crops, and maintain wildlife. Expanding human populations also put a strain on limited water resources.

3.10 FIRES

3.10.1 Characteristics and examples

Naturally occurring wildfires are relatively uncommon on islands compared to continents. Most natural fires are triggered by lightning but sometimes by volcanic activity such as lava flows or explosive eruptions. Strong, hurricane-force winds can spread fires even during heavy rainfall. Fires vary in their intensities, depending on their temperature, duration, and location within the canopy or soil. Slow-burning peat fires can last for weeks as the organic-rich soil gradually burns, whereas a canopy fire can pass so quickly that many understory plants remain unburned.

Fire has been of variable importance to the ecology of the nine island groups. Iceland rarely burns except in severe droughts accompanied by northerly winds when grasses can burn. The British Isles are also not particularly fire-prone. Lightning-triggered fires may have been more common 10 000 years ago when the climate was probably warmer and drier that it is today. For both the British Isles and Japan with their relatively long human history, it is difficult to sort out natural from human-related fires (see Chapter 5). Today about 1% of forest fires in Japan (about 20–25 per year) are caused by lightning.

On the tropical islands of Tonga, Puerto Rico, and Jamaica (Fig. 3.18), wildfires vary in importance, with dry, leeward areas the most fire-prone. Even wet tropical forests on the windward sides of these islands can burn during prolonged droughts, but most fires on these islands are triggered by lightning in the dry forests, arid shrublands, and grasslands on the leeward side. In Hawai'i, there are records of pre-human fires on O'ahu in both wet and dry lowlands and on the volcanic slopes of Mauna Loa volcano on the Island of Hawai'i. However, because native vegetation is not flammable, fire-dependent, or fire-maintained it is likely that fire played a limited role on these islands, even in the drier ecosystems.

Fires played a small role in the pre-human history of New Zealand of the eastern, drier regions of New Zealand. There is evidence of periodic, localized loss of forests by fire and replacement by ferns and grasses. Return times of fire, most likely caused by lightning, were apparently at least 1000 years, allowing reestablishment of forests before the next fire. The fire regime changed drastically in extent and frequency after human settlement (see Chapter 5). The Canary Islands have a mostly Mediterranean climate with frequent, low-intensity fires

Fig. 3.18   Use of fire to clear weeds for agriculture on steep, unterraced hillsides at St. Helen's Gap at the head of the Green River Valley, Blue Mountains, Jamaica (see Box 7.1).

(especially during drought years). The high frequency of fires means that there is little to burn each time.

3.10.2 Ecological effects and responses

Animals flee fires when they can, but then return as forage plants recover. Fires kill plants that are not fire-adapted, so an occasional, intense fire can reset the process of vegetation recovery, perhaps to a more fire-adapted flora. Other species may not be killed but regrowth from sprouts might be very slow in the post-fire conditions. Fire-maintained species benefit from fire effects because they utilize the increased nutrient levels from the ashes, increased light, and decreased competition following the opening of the canopy after a fire. Other species are actually fire-dependent. Cones of some conifers such as pines found on the Canary Islands require the heat of a fire in order to release their seeds from cones (serotiny). Fire has certainly been an important factor shaping the evolution of plant communities, particularly in the Mediterranean-like climates (mild winter with a hot, long summer and low overall precipitation) found on the Canary Islands.

3.11 ANIMAL ACTIVITIES

Animals on islands differ from mainland animals in many ways because of their isolation and evolutionary history (see Chapter 4). For example, islands often have more seabirds, both in terms of abundance and number of species, but fewer mammals than on mainlands, or no native mammal species at all. However, on both islands and mainlands, animals are agents of disturbance when their activities reduce biomass or alter ecosystem structure in relatively sudden ways. For instance, insects eat plants all the time, so herbivory is a part of the ecosystem. When there is a sudden outbreak of bark beetles or leaf eaters such as locusts (see Section 5.5), plants over relatively large areas suffer damage and mortality; herbivory can then be considered a disturbance. Animals also are affected by disturbances and have their own set of species-specific responses to each disturbance. The beetles that increase in population because of increased dead plant material following an outbreak of herbivores are responding to a disturbance caused by another animal species. An animal-induced disturbance has a characteristic intensity and frequency like any other disturbance.

Animals are agents of disturbance because of a wide range of activities. Grazers trample vegetation and compact the soil while herbivore outbreaks can defoliate whole forests. Animals that dig or burrow, such as sooty shearwaters, disrupt plant growth but aerate the soil so that some plants are found mostly on mounds related to burrow digging. Colonial animals such as nesting seabirds on isolated islands can have a disproportionate impact on their local environment (see Chapter 4). Seabird colonies alter nutrient cycles and dramatically influence the local vegetation (sometimes eliminating it entirely, sometimes speeding up succession in otherwise barren areas). For example, on Whakaari (White Island) in New Zealand, colonies of Australasian gannets have created barren patches of ground, damaged stands of forests, and have even buried markers used to designate study plots (Fig. 3.19). In contrast, on Surtsey, seabirds have provided missing nutrients and plant material from surrounding islands and the vegetation in their nesting areas is much more verdant than elsewhere. Finally, invasive animals can also create disturbances through their

Fig. 3.19   Effects of gannet colonies on Whakaari (White Island), New
Zealand include the destruction of vegetation and burial of study plot
markers.

exploitation of new habitats, introduction of novel activities or competitive superiority over natives. All islands are colonized by invading
species but humans have greatly accelerated the mixing of the world's
plants and animals while also driving some native grazing animals to
extinction, such as flightless birds in New Zealand (moa) and Hawai'i
(ducks and geese, see Chapters 5–7).

3.12 SUMMARY

Natural disturbances are a normal part of every island ecosystem. One
way of comparing the relative importance of disturbances across the
island groups is to consider the risk of human exposure to some of the
disturbances we have discussed (Table 3.5). Because of frequent earthquakes and cyclones, Japan and Tonga are clearly challenging places
for humans to live, even without considering volcanoes. The Caribbean
islands are also impacted, particularly by cyclones. However, even
island groups such as Iceland and the British Isles are not immune
from occasionally devastating disturbances, as we have seen.

Understanding how disturbances influence organisms is critical
to understanding how the arrival of humans has altered this interaction of the physical and biological worlds. Disturbances are best

Table 3.5. *Risks of exposure to natural disturbances among the nine island groups (percentage of population exposed per year). Ranks were determined from the sum of percentages in each column, where 1 = lowest risk, 7 = highest risk. Comparable data were not available for the Canary Islands or Hawai'i. Zeroes indicate negligible risk*

Island group	Overall Rank	Earthquake	Landslide	Cyclone	Flood	Drought	Tsunami
Iceland	1	0.5	<0.1	0	<0.1	0.5	0
British Isles	2	<0.1	<0.1	0	0.5	5	<0.1
New Zealand	3	2	<0.1	0.5	1	2	2.5
Jamaica	4	1.5	<0.1	10	0	2	0
Puerto Rico	5	0.5	<0.1	9.5	0.5	2	1.5
Tonga	6	22	0	10	<0.1	<0.1	<0.1
Japan	7	10.5	<0.1	18	<0.1	3	3.5

Source: preventionweb.net/countries

examined as a disturbance regime, or the complex set of interacting disturbances that impact a given site. Among the nine physical disturbances that we have discussed, there are many causal links (Fig. 3.20). Volcanoes can be triggered by earthquakes (and vice versa) and in turn cause tsunamis and other floods, landslides, fires, and aggradation of sediments into dunes. Earthquakes can be triggered by massive floods or landslides (and vice versa) and, in turn, can trigger fires and tsunamis. Each of these physical disturbances can potentially influence the disruptive activities of animals. To understand the effects of a given disturbance, it is critical to understand the context of the entire disturbance regime. Humans brought a new set of disturbances to the islands that they colonized (such as mining, agriculture, and invasive plants and animals) and influenced most of the existing physical disturbances in complex ways. Subsequent chapters address how these new disturbance regimes impacted natural ecosystems.

Tectonic disturbances are prevalent on all nine island groups. Volcanoes are more common in Japan and Iceland than elsewhere, except perhaps in such volcanic hotspots as Kamchatka (Russia), Java, and Ethiopia. Volcanic eruptions are hard to escape when the lava flows to the ocean and the ash and fumes cover the land. Therefore,

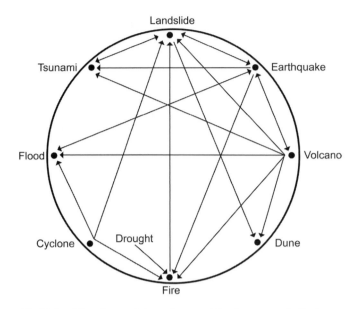

Fig 3.20. Disturbance interactions on islands; arrows indicate direction of influence between disturbances; multiple disturbances present at a given site constitute a disturbance regime.

island volcanoes are particularly destructive to life. Earthquakes can have the same impact as a large volcano, leaving little opportunity for escape. Volcanic debris and earthquakes, as well as terrestrial and submarine landslides trigger tsunamis, a disturbance to which islands are particularly vulnerable because of their high coast to area ratios. Small, low islands can be completely inundated by tsunamis.

Islands are also particularly vulnerable to cyclones. Cyclones usually diminish when they move inland but on small islands their winds do not diminish because they retain their energy source from warm ocean water. Cyclones, like tsunamis, also approach islands from unpredictable directions, making preparation for them difficult.

Drought and fire devastate islands that have minimal and ephemeral fresh water resources. The topography of the island groups that we cover ensures that they have significant precipitation that maintains aquifers, but these sources are quickly curtailed if there is less than average rainfall for even a few weeks. Therefore, flooding is not as serious a disturbance on islands as on mainlands with larger catchment basins. Aggradation, through movement of sand in dunes or from other particles, is similarly less important on islands than mainlands that have larger fetches to accumulate sediments. However, islands still receive many windblown particles from such places as the Sahara Desert to the Canary and Caribbean Islands or from the Gobi Desert in China to Japan and Hawai'i. Finally, the effects of seabirds that preferentially colonize small, remote islands are a reminder that island fauna have an important part in the local disturbance regime.

In the next chapter, we address how the remoteness of islands influences the response of animals and plants to disturbance. We discuss how organisms colonize new islands and how they evolve into the wondrous array of organisms that makes island biology clearly distinct from that of mainlands. Finally, we address the unique aspects of extinction on islands, in part due to the natural disturbances discussed in this chapter, but accelerated by the appearance of humans.

SELECTED READING

Allen, R.B., P.J. Bellingham and S.K. Wiser (1999). Immediate damage by an earthquake to a temperate montane forest. *Ecology* **80**: 708–714.

Barker, D. and D.F.M. McGregor (1995). *Environment and Development in the Caribbean: Geographical Perspectives*. Kingston, Jamaica: The Press, University of the West Indies.

Bellingham, P.J, E.V.J. Tanner and J.R. Healey (1995). Damage and responsiveness of Jamaican montane tree species after disturbance by a hurricane. *Ecology* **76**: 2562–2580.

Bryant, E. (2001). *Tsunami: The Underrated Hazard*. Cambridge: Cambridge University Press.

Davis, L. (2002). *Natural Disasters*. New York: Checkmark Books.

del Moral, R. and L.R. Walker (2007). *Environmental Disasters, Natural Recovery and Human Responses*. Cambridge: Cambridge University Press.

Elsner, J.B. and A.B. Kara (1999). *Hurricanes of the North Atlantic: Climate and Society*. Oxford: Oxford University Press.

Johnson, E.A. and K. Miyanishi (2007). *Plant Disturbance Ecology: The Process and the Response*. New York: Academic Press.

Martínez, M.L. and N.P. Psuty (2008). *Coastal Dunes: Ecology and Conservation*. New York: Springer.

Mosley, M.P. and C.P. Pearson (1997). *Floods and Droughts: The New Zealand Experience*. Wellington: The New Zealand Hydrological Society.

Nott, J. (2006). *Extreme Events: A Physical Reconstruction and Risk Assessment*. Cambridge: Cambridge University Press.

Ogden, J., L. Basher and M. McGlone (1998). Fire, forest regeneration and links with early human habitation: evidence from New Zealand. *Annals of Botany* **81**: 687–696.

Packham, J.R. and A.J. Willis (1997). *Ecology of Dunes, Salt Marsh and Shingle*. New York: Springer.

Restrepo, C., L.R. Walker, A.B. Shiels, *et al.* (2009). Landsliding and its multi-scale influence on mountainscapes. *BioScience* **59**: 685–689.

Sidle, R.C. and H. Ochiai (2006). *Landslides: Processes, Prediction, and Land Use*. Washington, DC: American Geophysical Union.

Stone, C.P. , C.W. Smith and J.T. Tunison (eds.) (1992). *Alien Plant Invasions in Native Ecosystems of Hawai'i: Management and Research*. Honolulu, HI: University of Hawai'i Press.

USGS Earthquakes Hazard Program (2009). http://earthquake.usgs.gov/regional/world/index.php?region=New%20Zealand (accessed 2 July 2009).

Walker, L.R. (ed.) (1999). *Ecosystems of Disturbed Ground*. Ecosystems of the World Volume 16. Amsterdam: Elsevier.

Walker, L.R. and R. del Moral (2003). *Primary Succession and Ecosystem Rehabilitation*. Cambridge: Cambridge University Press.

4

The plants and animals of islands

4.1 INTRODUCTION

Islands and their plants and animals command a place in the imagination. A cartoonist's image of a single coconut tree on an atoll instantly conjures an imaginary world of exile and self-sufficiency. Some island plants have fascinated people because of the myths that surround them, others for their size, and others for their rarity. A palm, coco-de-mer, found on two small islands in the Seychelles in the Indian Ocean, inspired erotic myths because of its voluptuous fruit, which are the largest and heaviest (up to 42 kg) of any plant. Until their native habitat was discovered in the late 1700s, all that was known of the plant was its fruits found floating at sea; these were imagined to have come from trees growing on the sea floor. Mythical large animals from islands also could have had some factual basis; for example, the giant roc, a bird that terrified Sinbad, could have been inspired by encounters with the now-extinct elephant bird of Madagascar.

The plants and animals found on islands have also provided economic wealth. Several species of frankincense trees native to the island of Socotra in the Indian Ocean contributed substantially to the wealth of Southern Arabian kingdoms 2000–3000 years ago. The incense derived from the resin of these trees was traded to the Egyptians, the Greeks, and the Romans who used it medicinally, as personal fragrances, and for religious and funeral rites. In the nineteenth century, the various species of sandalwood tree native to Hawai'i provided a short-lived economic boom for the Hawaiian aristocracy. The trees were felled for export to China, where they were made into incense. Abundant seabirds have deposited guano for thousands of years on some tropical islands such as Nauru, in the central

Pacific Ocean. Over time the guano has hardened and compacted to become rock phosphate. Rock phosphate on Nauru was mined, processed, and distributed as fertilizer for the pastures of New Zealand, which contributed to New Zealanders' prosperity in the 1960s (see Chapter 7).

The unique plants and animals of islands were critical in providing insight into the origin of species. Alfred Wallace's extensive exploration of the fauna of the islands of modern-day Indonesia and Malaysia in the 1850s gave him such insights and he further observed that the degree of isolation of an island and the geological age of an island were likely to determine how distinct were its plants and animals. Charles Darwin's formulation of evolutionary theory was influenced by his correspondence with Alfred Wallace and his own experiences of islands' plants and animals. Darwin's collection of mockingbirds from different islands of the Galápagos archipelago was examined by John Gould, who noted that there were subtle differences between the birds from each island. Darwin proposed that these differences were adaptations to the variable environments of each island.

Since Darwin's and Wallace's vital first observations, the discipline of understanding the plants and animals of islands and how they came to exist has developed further and is now known as island biogeography. The name of the discipline reflects Wallace's realization that a knowledge of the biology of the plants and animals, the geography of the islands that they colonize, and the inter-relationships between these are all needed to explain current patterns. In this chapter, we consider the origins of the plants and animals from the nine island groups and their often unusual characteristics. We conclude by examining features of island plants and animals that have resulted in disproportionately high rates of past extinctions and likely will lead to much future extinction.

4.2 HOW ISLANDS GAIN THEIR PLANTS AND ANIMALS

The plants and animals we see on islands today derive from organisms that either dispersed (often long distances) over the sea or across historical land connections to other land masses. Both of these processes have contributed to the composition of the native plant and animal communities on islands. When volcanic islands form far from other land masses they gain their plants and animals entirely by long-distance dispersal.

4.2.1 Dispersal

A new island

Surtsey, off the south coast of Iceland, is one of the world's youngest islands. The island rose out of the North Atlantic Ocean to form land in 1963 (see Box 2.2). New islands like Surtsey give insight into how plants disperse to and colonize an island and how certain species of plants become dominant, while other species remain scarce. A new volcanic surface is a harsh environment: some key nutrients essential for plant growth, such as nitrogen, are very scarce in the new soils. Not all plants that are dispersed to Surtsey are likely to survive. Annual surveys of Surtsey recorded 69 plant species that had colonized between 1963 and 2008. All of these probably derived from Iceland or nearby islets. Of 32 plant species that had established persistent populations, 9% had probably floated to the island across the sea (Fig. 4.1), 16% had probably been blown to the island by wind, and 75% were likely to have been carried to the island by birds, especially by gulls. One of the gull species that nest on Surtsey, the lesser blackbacked gull, is not only the most likely agent of dispersal for plants, but it also favors a particular set of species. These gulls feed mostly in meadows on Iceland and adjacent islets and therefore are most likely to bring seeds to Surtsey from these habitats. Furthermore, they nest colonially in a single area on Surtsey, limiting the area to which the seeds are dispersed. Guano deposited by the birds in these habitats has high nutrient concentrations that favor several perennial grass species (see Section 3.11 and Fig. 3.19). Lower nutrient sites outside the gull colony have lower plant cover and are dominated by the herbaceous sea sandwort. Plant colonization processes on Surtsey therefore show us that dispersal over sea can occur by multiple means and that even in the very early stages of an island acquiring a new community of plants and animals, certain species will be favored more than others.

Dispersal over sea

The plants and animals found on Iceland, which provided the source of Surtsey's colonists, are obviously of older origin, but compared with many islands, they too have colonized quite recently and all by dispersal over sea. The surface of Iceland was laid bare by repeated severe glaciations from about 1.6 million years ago, with most of the last glaciers retreating only about 10 000 years ago. It is conceivable that

Fig. 4.1   Surtsey, a new volcanic island that formed south of Iceland in the 1960s. Initial colonization was dominated by circular patches of sea sandwort, a water-dispersed plant.

some plants and animals may have survived glacial advances around volcanic hot springs that provided relatively warm refuges. After the retreat of most of the glaciers, plants and animals re-invaded and colonized Iceland almost entirely from northwestern Europe, even though Iceland is about 1300 km from Norway. Among the means of dispersal of some animals may have been fresh melt water from the mainland mixing with ocean surface currents in the North Atlantic as the ice sheets melted. Ocean currents have also been invoked as a mechanism to explain how a heavy, flightless, soil-bound weevil colonized Iceland. The same species occurs in northwestern Europe and also on small, remote, North Atlantic islands such as the Faroe Islands.

The islands of Hawai'i are even further away from the nearest land masses than Iceland (see Chapter 2). All of the plants and animals native to these islands arrived by long-distance dispersal. Recent analyses of the DNA in Hawai'i's plant and animal species are yielding insights not only into the ancestors of some of the islands' unique plants and animals but also some unexpected sources from which the ancestors colonized.

Many of the Hawaiian plants and animals colonized from the western Pacific and evolved and diversified subsequently. Hawai'i has an extraordinary diversity of *Drosophila* vinegar flies – about 1000 species. Vinegar flies probably arrived from the Western Pacific about ten million years ago onto an island west of Kaua'i that has now submerged below the Pacific Ocean. From there, vinegar flies colonized the younger islands to the east as they emerged. 'Ohia is a red-flowered tree in the myrtle family and is the most abundant tree in Hawai'i's rainforests. Its ancestors also arrived by long-distance dispersal from the west but probably more recently than the vinegar flies. 'Ohia is very closely related to trees in New Zealand and has close relatives on many other remote Pacific islands such as Tahiti. Seeds of all these trees are tiny, light, and wind-dispersed and it is likely that a New Zealand ancestor is the ultimate source of most of the related species of remote Pacific islands including Hawai'i. 'Ohia or its ancestor is most likely to have arrived first on Kaua'i and spread to the younger islands from there. Its likely time of arrival in Hawai'i is disputed – one estimate of its arrival is less than 2 million years ago but more recent estimates are much older; 4–6 million years ago or more.

In contrast to settlement patterns from the west, two other groups of plants are likely to have colonized Hawai'i by long-distance dispersal from the east. This dispersal pattern is remarkable because of the long distance involved (3820 km to mainland USA) and almost no islands on which to "island-hop." Silverswords are spectacular, large herbs in the daisy family characteristic of Hawai'i's mountain vegetation (Box 4.1, Fig. 4.2, and Plate 7). Their ancestor is most likely to have been a rather drab tarweed dispersed from western North America that arrived about five million years ago. Western North America is also a likely origin for Hawaiian sanicles, small herbs in the carrot family, which were probably dispersed by adhering to birds' feathers; modern North American relatives of the sanicle have hooked prickles on their fruits.

Box 4.1. A rare but beautiful plant

Silverswords are well-known for their beautiful flower stalks and silvery leaves. They often grace the cover of natural history and conservation magazines and are featured on television because

Box 4.1. (cont.)

they are so visually appealing. Their animal analogy might be the panda. Part of a larger radiation of 30 plant species (the silversword alliance) that radiated from a western North American tarweed, silverswords add their vibrant colors to the alpine vegetation of volcanoes on Maui and the Island of Hawai'i. I was introduced to these wonders by my fiancée, getting romance and botany in a two-for-one package! But access to the Mauna Kea silversword (Fig. 4.2) on the Island of Hawai'i was along a rough 10 km road that took an hour to drive, followed by scrambling up steep canyon walls to find the few remnant natural plants that had survived extensive grazing by introduced mouflon sheep and goats. Luckily, seeds of these survivors had been saved and restoration is underway. When the rosette of leaves of a silversword reaches about 30 cm in diameter, a showy flower stalk grows to several meters in height, probably attracting pollinators from long distances. Native bees and moths pollinate the flowers and then the rosette dies. The silver hairs on the leaves may reduce water loss in its dry alpine habitat and gave the plant its Hawaiian name, 'ahinahina or "very silver." LRW

Remote islands can be viewed as the end of the line for long-distance dispersal, but this need not be the case. Monarch flycatchers are birds that occur on small central Pacific islands, including Hawai'i, and their ancestors colonized these islands by long-distance dispersal from eastern Asia more than 1.5 million years ago. Subsequent colonization of the rest of the Pacific, with 55 modern species from Australia to Tonga to the Marquesas Islands probably derives from ancestors that dispersed from small central Pacific islands. Remote islands like Hawai'i can also be sources for plant dispersal. The most likely origin of the *Melicope* trees in the Marquesas Islands involves at least one instance of long-distance dispersal from Hawai'i, with seeds probably transported by a migrating bird.

Dispersal along land bridges and by floating

Jamaica and Puerto Rico are continental fragments (see Chapter 2). Although land bridge connections are likely in their geological past, dispersal over water is the most likely origin of most of the native plants and animals on both islands today. Both islands were largely submerged

Fig. 4.2   The Mauna Kea, Hawai'i silversword, an endangered plant growing only on the Island of Hawai'i in areas protected from grazing by feral mouflon sheep (see Box 4.1).

for millions of years. No part of Jamaica was more than a few meters above sea level between about 15 and 44 million years ago. Before then at least three terrestrial animals, including a primate, occurred in Jamaica. These animals have been found as fossils in western Jamaica (between about 44 and 55 million years old) and they probably colonized when this part of the island was either connected to or close to Central America. However, it is most likely that animals that had colonized the island over land bridges became extinct when the island was subsequently drowned or was fragmented into smaller islets. Similarly, most of Puerto Rico was submerged 5–35 million years ago, although it is likely that some land remained exposed throughout. After being largely submerged, Jamaica and Puerto Rico emerged from the sea, about 15 million and 5 million years ago, respectively, and formed new landscapes but were not again connected to larger land masses. With new land to colonize, most of the ancestors of today's native Jamaican and Puerto Rican plants and animals did so by dispersing over water. Cuba was an important source of colonizers to Jamaica, and Hispaniola an important source for Puerto Rico. Plants and animals dispersed along the West Indian island chain, and provided a link between South and North America. For example, it is likely that the family of birds that includes orioles and cowbirds dispersed from North America to South America through the islands of the West Indies before the formation of the Panama Isthmus, between 3 and 3.5 million years ago.

Land animals native to Puerto Rico and Jamaica include endemic species of boa, and Jamaica had an endemic species of monkey that became extinct after human settlement. Jamaica has an endemic species of iguana as does the island of Mona, located between Puerto Rico and the Dominican Republic. Over-water dispersal as a means of colonization by such large, non-swimming land animals as monkeys, boas, and iguanas may seem improbable but it can occur in extraordinary circumstances. In late 1995, Hurricane Luis passed through the eastern Caribbean and, in its aftermath, at least 15 green iguanas were washed on a mat of logs and uprooted trees onto the beaches of the small island of Anguilla. These iguanas did not previously occur on the island but established and were breeding on the island three years later. It is likely that the iguanas were dispersed from the larger island of Guadeloupe, 320 km to the south.

Like Jamaica and Puerto Rico, New Zealand's origins are continental. The New Zealand islands began separating from the large land mass of Gondwana about 80 million years ago. Some of the unique plants and animals of modern New Zealand almost certainly derive from ancestors that were on the land mass that separated. It is possible that even the extinct species of flightless moa may have colonized New Zealand after its separation from Gondwana, about 60 million years ago. If this is the case, the moa ancestor, most closely related to the modern tinamous of South America, flew to New Zealand and the ancestral moa lost the power of flight during the next 3 million years. There was a period during which much of New Zealand's land mass was submerged (about 35 million years ago) and the small part of the land that was not submerged was a chain of islands. The landscape that emerged after this period of submergence presented new opportunities for colonization. Most of today's New Zealand native plants and animals derive from ancestors dispersed over water. The importance of long-distance dispersal can be seen even over the short history of human settlement in New Zealand. Birds such as silvereyes, welcome swallows, and white-faced herons that are now among New Zealand's most abundant birds, have colonized over the Tasman Sea from Australia within only the last 150 years. For many species, New Zealand is a transit point for further dispersal into the Pacific. The Chatham Islands, 800 km to the east of the South Island, emerged from the Pacific Ocean between 1 and 3 million years ago and all of the plants and animals of those islands, most of which are endemic and which include flightless insects, almost certainly derived from ancestors dispersed over water from New Zealand (Plate 8). We can observe the process of dispersal to New Zealand's remote islands even over short periods with species introduced by Europeans. For example, European starlings have dispersed over water from the main islands of New Zealand, to which they were introduced, and founded populations on all of its remote islands, from Raoul Island 1000 km to the north to Campbell Island 600 km to the south of the main islands.

Dispersal among islands

Dispersal within island groups and over short distances is also important in explaining why certain plants and animals are present on some islands and not others. This can be linked to interactions between plants and animals. For example, pigeons and bats play a key role in dispersing the seeds of some plants over short distances over sea to many tropical and some temperate islands. Several Tongan trees, especially those with large seeds, probably depend on fruit bats and pigeons to colonize recent volcanoes in the group. If pigeons or fruit bats become extinct or are reduced in numbers, this reduction can determine the kinds of forests that develop, favoring species that do not require these dispersers. Small populations of plants endemic to islands can become restricted in range if their dispersers are not present. For example, the tree *Elingamita johnsonii* is presently restricted to two tiny islets in the Three Kings Islands, 60 km northwest from the northern tip of the North Island of New Zealand. Only the native New Zealand pigeon swallows its fruit and thus disperses its fruit but these birds have been absent from the islands for at least 100 years; the largest nearby island in the archipelago, which could have supported a population of these pigeons, was severely deforested and grazed by goats. Goats were eradicated in 1946 and since then the forest has regenerated. Pigeons have recolonized only within the last decade but it is still too early to tell if they will disperse the rare *Elingamita* tree to other islands and expand its range.

4.2.2 Past land connections

Islands on continental shelves, such as the British Isles, may have acquired some of their species by dispersal over sea but it is evident that most of the plants and animals of even the remote offshore islands of the modern British Isles derive from its former connections to continental Europe. So severe was glaciation of the British landscape during the Pleistocene (much of the past 1.6 million years) that it laid bare the landscape for fresh colonization when the glaciers retreated. With retreating glaciers came rising sea levels, thus colonization over land could be halted if a species did not disperse quickly enough to beat rising sea levels. This is why, for example, Ireland lacks snakes.

The plants and animals of the main islands of Japan, a continental fragment, derive mostly from a series of past land bridges, which formed and vanished largely in response to sea level fluctuations due to water being locked up in ice caps. Although the Sea of Japan opened to begin Japan's isolation between 16 and 25 million years ago, there have been connections that enabled repeated colonizations from Asia at various times since, especially during the Pleistocene. The most recent connections are to the north so that there were routes for overland immigration to the northern boreal island of Hokkaido until less than 10000 years ago. Hokkaido was separated from the other three main islands of Japan (Honshu, Shikoku, and Kyushu) by the deep waters of the Tsugaru Strait (Fig. 4.3). Temporary ice bridges during the period of maximum advance of glaciers or of sea ice may have allowed some mammals to colonize south from Hokkaido and vice versa but there was no land bridge to cross. Honshu, Shikoku, and Kyushu have been interconnected by land bridges at various times, especially during glacial advances, so they share many species. These three islands have been connected in turn by major land bridges to eastern Asia. For example, a land bridge to northern China that existed between 450000 and 600000 years ago was the likely route by which now-extinct straight-tusked elephants colonized Japan. The most recent land bridge connections to Asia during glacial advances may be as recent as 300000 years ago. Multiple invasions across land bridges to Japan could also explain differentiation in some Japanese animals. For example, genetically distinct populations of wild boars may have invaded across land bridges to Japan at different times, and subsequent mixing of these produced the modern Japanese wild boars.

South of the main large islands of Japan, the Ryukyu archipelago extends from south of Kyushu almost as far as Taiwan. The

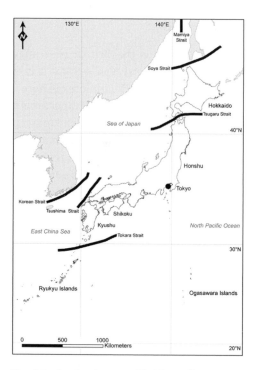

Fig. 4.3   Straits that provided impediments to animal and plant colonization of Japan, especially the deep Tsugaru and Tokara Straits.

plants and animals on these islands derive from two different land bridge connections. The first land bridge extended from Japan's southern main island, Kyushu, for about 800 km south to the deepwater Tokara Strait (Fig. 4.3). This was the principal route for colonization of the northern part of the Ryukyu archipelago by land mammals and plants. Connections between these islands and Kyushu ended around 500 000 years ago. Since then, distinct plants and animals have developed on individual islands. The second land bridge extended north from Taiwan and includes such islands as Okinawa and allowed colonization by the ancestors of distinctive animals, including some confined to one or two of these islands, such as the Amami rabbit and the Iriomote cat. Hence, animals in the northern Ryukyu Islands have ancestors of arctic origins, while those in the southern Ryukyu Islands have subtropical origins. The result is distinct differences in the communities of plants and animals found on islands less than 100 km apart on either side of the Tokara Strait. In contrast to the Ryukyu archipelago, the Ogasawara (or Bonin) Islands, which extend south

Fig. 4.4   Tuatara on Takapourewa (Stephens Island), Cook Strait, New Zealand.

from Tokyo, are oceanic, volcanic islands and lack any land bridge connections. The ancestors of their plants and animals arrived by long-distance dispersal.

We have already emphasized the role of long-distance dispersal in explaining the composition of New Zealand's current flora and fauna. However, the most likely explanation for the origins of some of its most distinctive animals is not long-distance dispersal but that their ancestors were on the part of Gondwana that separated to become New Zealand. These include the tuatara, a relict from an order of reptiles that otherwise became extinct 145 million years ago (Fig. 4.4); the *Leiopelma* frogs, which are among the most primitive of frogs, with their closest relatives in North America; and the New Zealand wrens. The ancestors of all of these are likely to have survived after the separation from Gondwana on small islands when most of the New Zealand land mass was submerged 35 million years ago. Even after New Zealand separated from Gondwana, it is possible that land bridges to the north, or at least chains of islands along now submerged seamounts, enabled dispersal among such remote points as New Zealand, Lord Howe Island, Norfolk Island, and New Caledonia. Such explanations are invoked to explain why a species of carnivorous *Placostylus* snails on the Three Kings Islands off the northern tip of New Zealand are more closely related to those on New Caledonia, over 1500 km away, than they are to the species on the New Zealand mainland that are only 60 km away.

4.3 EVOLUTION OF NEW SPECIES

Once plants or animals disperse to an island and establish breeding populations, or when populations are marooned on an island after a land bridge submerges, a number of interrelated factors determine whether those populations evolve over time to form new, distinct species.

4.3.1 Time

The age of an island determines whether it is more likely to have unique species of plants or animals. On an island as young as Surtsey, populations of plants that are less than 50 years old and with few generations are hardly likely to differ from the populations from which they derive. However, there may be selection for different attributes in a newly established plant population on an island. The very attribute of a wind-dispersed plant that enabled it to colonize an island could be detrimental to it establishing a permanent population on the island: its seeds could be blown into the ocean rather than remain on the island. Furthermore, the seeds that arrive on an island are most likely to derive from the subset of a mainland population that produces seeds best able to travel long distances and this attribute will be disproportionately part of the gene pool of a newly established population. Natural selection would favor offspring that disperse seeds on the island to build populations and there is evidence for this. Populations of two different short-lived herbs in the daisy family showed just this response on islands off the coast of Vancouver Island in Canada within a decade of establishment. The longer a population had been established on an island, the more the seeds of the plants in the population had greater volume than seeds of plants of newly established populations on other islands or on Vancouver Island. Plants in the daisy family also have a tuft of hairs attached to the seed, a pappus, that enables seeds to be blown by the wind, and the longer populations had been established on islands the more the pappus reduced in volume. The combined effect of increasing seed volume while reducing pappus volume resulted in seeds of the island populations being less suited to disperse over long distances. Evolving larger seeds is not only about reducing dispersal ability in species that could be blown off islands. Reduced competition and the absence of specialized herbivores or seed predators compared with the mainland enables island plants to invest more energy in seeds. Thus, not only wind-dispersed plants but also animal-dispersed plants show a tendency to develop larger fruit than their mainland ancestors.

The length of time an island has been isolated determines the likelihood that new species will evolve on it. We can see examples of this on small islands 30–60 km from the northern coast of the North Island of New Zealand, all formerly connected by land bridges. The number of woody plant species found today on these islands reflects the time since land bridges ceased to exist. The Three Kings Islands, which were last connected by a land bridge 6 million years ago, have 12 endemic woody plant species. The Mokohinau Islands, last connected when sea levels rose after the last glaciations about 12 000 years ago, have only one endemic woody plant.

4.3.2 Isolation

The more remote an island is the less frequently colonizing individuals will reach it. Should they colonize and establish breeding populations, these new populations will develop from the pool of genes of their founders. Along a gradient of isolation for the island groups that we consider, we can see that the most isolated islands often have the greatest level of endemism – species unique to them (Table 4.1). The British Isles, close to continental Europe, and Japan, close to eastern Asia, have low endemism in their native plants. Although geologically similar to the islands of Hawai'i, the Canary Islands, which are close to northwestern Africa, have much lower endemism than very remote Hawai'i. Iceland, which is remote, clearly breaks the pattern with no endemism in its flora. All of Iceland's plants have probably re-colonized the island within the last 100,000 years because before that time the island was almost totally denuded by advancing glaciers. Thus, isolation needs to be interpreted as a mechanism for the rate of species formation as well as the time that an island has existed and its history of disturbance.

4.3.3 Size and topography

The larger an island is, the more species it can contain. This relationship has been shown for plants and animals in many parts of the world, both within and among island groups. For example, the number of native plants on islands in the West Indies, including Puerto Rico and Jamaica, show a clear relationship with island size. In cases where this relationship is weak, recent extinctions may be the explanation.

Table 4.1. *Native and endemic higher plant species on each island group*

Island	Number of native plant species	Percent endemic (number of endemic plant species)
Iceland	470[a]	0 (0)
Tonga	533	3 (18)
British Isles	1 481	3 (48)
Japan	5 565	c. 4 (>222)
Puerto Rico	2 128	10 (215)
Canary Islands	1 860	28 (521)
Jamaica	2 746	31 (852)
New Zealand	2 362	82 (>1 930)
Hawai'i	1 009	90 (906)

Note: [a] native and introduced.

For example, recent extinctions are likely to be a reason that there is a poor relationship between the sizes of islands in Tonga and the number of birds currently on the islands. When the number of birds per island also includes extinct species known from fossil remains, the relationship between island size and number of bird species is stronger. The relationship can also be weak on other islands because of their history. For example, Iceland is a relatively large island but it has few plant species (Table 4.1) because of its climate and because of its recent post-glacial history.

In many cases, across a gradient of increasing island size, topographic variation also increases (see Chapter 2). Atolls in tropical oceans are an exception to this. Whether atolls are large or small, they usually are only a few meters above the sea and have little variation in topography. Therefore, size alone is likely to be a reason that a large atoll supports more species than a small one. For example, in islands northwest of Hawai'i, larger atolls have more plant species than smaller ones. Although large atolls can support more species, the number of plant and animal species unique to them is small. Indeed, many of the plants on atolls are in common across all tropical seas. For new species to evolve, more complex habitats are needed in which plants and animals have opportunities to colonize and develop populations in a range of new microenvironments. This process takes place in tectonically active regions such as Tonga, where reef islands (makatea) have been raised out of the ocean. These islands have new microenvironments. Soils develop from the dead former coral and from limestone on the ocean floor. The central lagoon of an atoll, once raised out of the ocean by tectonic movement, often becomes a freshwater wetland suitable for forest development, and the species colonizing such areas are those that can tolerate poor drainage. In extreme cases, tectonic movement can tilt a raised reef island, as is the case on the island of 'Eua in southern Tonga. The result is that the eastern side of 'Eua is now comprised of limestone cliffs and the island's summit is over 300 m above the sea (see Section 2.2.1). 'Eua can support a diversity of rainforest trees from sea level to the island's summit in a way that most raised reef islands of the same size cannot. This unusual topography also afforded opportunities for other species to develop. Fossils of birds excavated from limestone clefts and caves on 'Eua show that it formerly supported up to three bird species that may have been endemic to the island – a rail, a megapode, and a songbird. All of these became extinct after human settlement began on the island about 3000 years ago.

Volcanic oceanic islands are often topographically complex and greater complexity can result in strong environmental gradients. There are strong temperature gradients on elevational gradients from sea level to the summits of such high mountains as Mauna Kea (4207 m) in Hawai'i and Pico del Teide (3718 m) in the Canary Islands; their upper slopes provide alpine habitats. These volcanic mountains are also affected by trade winds so that there is significant rainfall on the windward sides of islands and rain shadows on the leeward side (see Chapter 2). As a result, these islands provide very diverse habitats, and single founding populations of species have evolved to form a diverse range of endemic species on such islands. Volcanic islands of lower elevation, such as Rarotonga or Tahiti in the eastern Pacific, are often cloud-capped and support rainforests in a way that nearby large atolls cannot. Tectonic movements that have produced axial mountain ranges on continental fragments, including Jamaica, Puerto Rico, Japan, and New Zealand, result in gradients of temperature and rainfall across these ranges. In all of these continental fragments, endemic species have evolved to occupy particular parts of these gradients. For example, *Portlandia* is a genus of shrubs with large flowers that is only found in Jamaica and the seven species all occur on limestone but on different parts of the island, some on the windward wetter side of the island and others on the leeward dry side (Fig. 4.5). An old, isolated island that is large and topographically diverse has a greater proportion of endemic species than a young, topographically homogeneous island.

Fig. 4.5   Flowers of *Portlandia grandifolia* in wet limestone forest on the Nassau Mountains, St. Elizabeth, Jamaica. All seven species of *Portlandia* occur naturally only in Jamaica.

4.3.4 Biological features

Founding populations of plants or animals on an island are likely to be small in number. Especially on very remote islands, there will be no or infrequent addition to the limited gene pool. These two factors are likely to accentuate particular features of the founding population's already limited gene pool. Furthermore, the limited area of islands, coupled with an even smaller area of suitable habitat for the plant or animal, further restricts population sizes and gene pools. Natural selection thus acts on small gene pools so that island populations can quickly differ from the populations they derive from and, in this way, new species arise.

When plants and animals colonize, the novel island environment can present major benefits compared with the original environment. Colonizers may escape specialized animal and plant diseases of their original ranges. New plant populations on many remote islands will experience no pressure from big grazing mammals. Grazing by big mammals selects for physical defenses in many plants. Freed from this pressure on islands, the energy expended in constructing defenses needed for protection from mammalian herbivores is no longer required and can be invested elsewhere. New, undefended species thus evolve. For example, palms often have spines or thorns on continents but seldom do on remote islands. Palms in New Zealand and Hawai'i and spurges (genus *Euphorbia*) in the Canary Islands (Fig. 4.6) do not

Fig. 4.6   *Euphorbia canariensis*, undefended by spines, near Punta Negra on the west coast of Tenerife, Canary Islands.

have spines or thorns. Similarly, predatory mammals seldom occur naturally on remote islands, and freed from pressure from predators, colonizing animal populations evolve behaviors and biological features, such as the loss of the ability to fly in birds, which result in new species. One result is that many of the bird species endemic to remote islands seem to have little or no fear of predators, including humans, such as the Hawaiian goose or nēnē (Fig. 4.7). Charles Darwin noted this on his encounters with endemic birds in the Galápagos archipelago. He predicted that when island species were faced with competitors or predators introduced by people from continents, they would be incapable of defending themselves.

Founding populations of species on an island may benefit from a low number of competing species or evolve a role played by others, thereby exploiting unused resources. On large, remote islands, the grazing and browsing roles filled by mammals on continents have been adopted by birds (New Zealand, Hawai'i) or tortoises (Galápagos,

Fig 4.7   The rare nēnē on the north coast of Kaua'i, Hawai'i. The
Canada goose is its likely ancestor.

Mauritius). Some of the New Zealand wētā, insects related to crickets, have adopted the role of seed predators and hoarders elsewhere
assumed by rodents (Plate 8).

Founding populations can derive from groups that are in a stage
of evolutionary radiation and thus are more likely to form new species
when isolated new populations develop, as on islands. For example,
Hebe is a genus of shrubs in New Zealand that has evolved into 88
species from a common herbaceous ancestor that arrived at most
5.5 million years ago (Fig. 4.8). On islands off the northern coast of the
North Island, which were connected by land bridges even as recently
as 12 000 years ago, there are three species or subspecies of *Hebe* that
are restricted to individual islands or small archipelagoes. Few other
unique species occur on these islands. Finally, another worldwide
driver of the rate at which species evolve is a gradient from the tropics to high latitudes. There is evidence that, within genera, the rate
of evolution of new species is most rapid in the tropics, the zone of
maximum solar energy, and declines toward the poles. This pattern
is likely to be a contributing factor to the fact that forests on tropical
islands have more endemic plants and animals than many temperate
islands, even when comparing islands of similar geological age.

Fig. 4.8   Alpine *Hebe* in flower in Fiordland, South Island, New Zealand.

4.4 SPECIAL FEATURES OF PLANT AND ANIMAL COMMUNITIES

4.4.1 Species richness

Islands have lower overall richness of species than the nearest continental areas. The more remote an island is, the more likely it is that the island or archipelago will lack groups of plants and animals present on the nearest mainland. For example, various evergreen oaks are dominant trees in rainforests in the mountains of Central America. The forests of the Blue Mountains of Jamaica and the Luquillo Mountains of Puerto Rico and other West Indian mountains lack oaks entirely. Oaks require mammals and particular birds that disperse the heavy seed. Therefore, they have never dispersed to Caribbean islands. Many large-bodied mammals in Central American rainforests, including predatory cats, and fruit-dispersing pigs and tapirs likewise have never been present in Jamaican and Puerto Rican forests. Major birds of Central American forests, such as toucans, have also never been present on Caribbean islands.

While islands are poor in numbers of species and often lack certain taxonomic groups of plants and animals, they can be the last refuge of plants and animals that have become extinct on continents. Competitors or predators that caused their extinction on continents did not reach islands. For example, todies are a family of small insectivorous birds related to kingfishers. The only surviving todies occur on the larger islands of the West Indies: five closely related species, with emerald green feathers and scarlet bills – one each on Jamaica, Puerto

Fig. 4.9   Montane rainforest interior near Mirador del Bailadero in the
Garajonay National Park, La Gomera, Canary Islands. The dominant
tree is *Erica arborea*.

Rico, and Cuba, and two on Hispaniola. Ancestors of todies probably
dispersed to the West Indies from North America during the Oligocene
(23–34 million years ago). Todies are known from only the fossil record
in North America and are also found as fossils in France and Germany.
We have already mentioned the extraordinary example of tuatara in
New Zealand, the only extant member of an order of reptiles that
became extinct elsewhere 165 million years ago.

The mild oceanic climates of islands can also be a reason for survival of entire ecosystems that disappeared on continents as a result
of climatic change. The Canary Islands, along with the Azores and
Madeira Islands, support temperate rainforests dominated by evergreen trees in the laurel family (Fig. 4.9). Forests composed of closely
related species were widespread in the Mediterranean Basin and as far
away as Hungary but became extinct in Europe, probably as a result
of the combined effects of aridification that resulted from the formation of the Mediterranean Sea and the Sahara Desert, and the effects
of repeated glaciations.

4.4.2 Diverse but related species

Charles Darwin's observations of the diversity of closely related birds
within and among different islands in the Galápagos archipelago highlighted that there can be extraordinary diversity among closely related

Fig. 4.10   *Eleutherodactylus* frog in montane rainforests, Blue Mountains, Jamaica.

species on islands. Modern studies of species, DNA reveal how these diversifications or radiations take place. These radiations of species, usually all endemic to an island or an island group, are one of the main reasons that some islands are international "hotspots" for biodiversity.

Several remarkable radiations of animals have taken place on Jamaica. Jamaica has 17 species of frog in the genus *Eleutherodactylus* (Figs. 4.10 and 4.11). All of these species evolved from a common ancestor, which probably arrived from Cuba about 22 million years ago. The modern species include ones that live in streams, rocky areas, caves, and within large bromeliad plants. Others are ground-dwelling, one lives in leaf litter, and another can climb the lower trunks of trees. Two of these frogs are widespread within Jamaica but most occupy narrow ranges. Eight of these frog species coexist in Jamaican rainforests in two parts of the island, on limestone in the west and in the upper parts of the Blue Mountains in the east, but only three species are shared between these two places. The land crabs of Jamaica offer another example of radiation. All nine species of land crabs in Jamaica derive from a marine ancestor that colonized the island only 4 million years ago. It is unusual worldwide for marine crabs to colonize the land, and those that colonized Jamaica have unique adaptations to life on an island including maternal care of juveniles. One of these crab species can complete its life history entirely in large bromeliad plants. The mother raises the young in the plant's leaf bases. Other species live in rock rubble, caves, or streams.

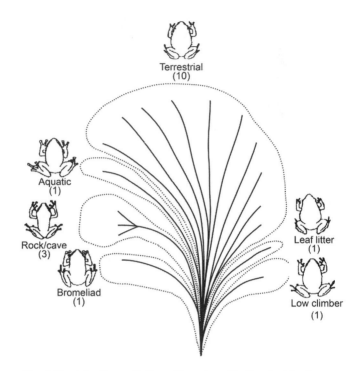

Fig. 4.11   Adaptive radiation of Jamaican *Eleutherodactylus* frogs showing how frogs found in different habitats differ in morphology. Numbers of species are in parentheses. Adapted from Hedges (1989).

On many islands, there are closely related species of plants that derive from a single ancestor. In the Canary Islands, there have been radiations of species in a genus of the borage family (*Echium*) and in relatives of the sow thistle (*Sonchus*). Both of these are represented by more than 20 species unique to the islands and occupy a wide range of habitats. All these species derive from original colonizers that were almost certainly herbs, but many of the new Canary Island species that evolved are woody. The woody *Sonchus* species derive from a single colonization, most likely on Gran Canaria, from which the other islands were colonized. New woody species evolved on all of them. Succulent species of the genus *Aeonium* in the stonecrop family show the greatest radiation with 36 species across the Canary Islands and up to 14 species on a single island; most are restricted to single islands. *Aeonium* also forms woody species. In contrast to *Echium* and *Sonchus*, *Aeonium* is likely to have evolved originally in the Canary Islands and has colonized other islands and Africa from them.

Fig. 4.12   A Hawaiian *Cyanea* in flower.

Nearly 10% of all of Hawai'i's native plants are endemic species of lobelias, and all are derived from a single immigrant species. These species are placed in six genera, of which the largest, *Cyanea*, has 79 species (Fig. 4.12). It is not yet clear when a common ancestor of *Cyanea* reached the islands of Hawai'i but it is likely to be 8–18 million years ago. It is probable that it first colonized an island older than Kaua'i which has now sunk below the sea, and dispersed to the younger islands from there. The modern species show that there have been several instances of dispersal and new colonization events among the islands. *Cyanea* has undergone a striking series of radiations in growth form, leaf size and shape, and flower shapes and sizes. There are species that are shrubs, trees, and even one vine-like species; most occur in wet forests (1000–2000 m above sea level). The different species of *Cyanea* may have played a role in the radiation of more recent arrivals – the endemic Hawaiian honeycreepers. The honeycreepers all derive from an eastern Asian finch that colonized Kaua'i about 3.5 million years ago. Over time, the ancestors of modern species dispersed to the other islands. There are 23 currently extant species of honeycreeper and at least 24 additional species that have become extinct since human settlement. Some of the extant species eat fruits and seeds, others eat mostly insects, and others mostly nectar. The nectar-feeders include species with long, curved beaks that match closely to the size and shape of flowers of particular *Cyanea* species and other lobelias that are unique to Hawai'i. The Hawaiian honeyeaters were also pollinators of *Cyanea* and the other lobelias. This family of birds was unique to Hawai'i and was most closely related to the silky flycatchers of North and Central America. Their ancestor probably arrived in the islands long before the honeycreepers' ancestor, between 14–17 million years ago, from either the Americas or north-east Asia. Unfortunately, this entire family, including five species that survived until historic times, is now extinct. The last, the Kaua'i 'ō'ō, became extinct in about 1987, probably a victim of introduced avian malaria.

4.4.3 Unusual life forms and behaviors

Flightlessness in birds, giant and dwarf forms of animals, and woodiness in plants that are usually herbs are among the unusual life forms that are peculiar to islands, especially those that are remote. We have already alluded to why some of these attributes might be favored on islands.

Rails are secretive ground or wetland birds and on many Pacific islands they are flightless; on other islands flightless rails have become extinct. Rails fly fairly frequently to islands in the Pacific; modern examples include buff-banded rails and their relatives, spotless crakes. On colonizing islands with no competitors, no predators, and no need to seek remote sources of food, there is less need to expend the energy required to maintain flight muscles. Under such conditions, natural selection no longer favors the ability to fly and new flightless species of rail have evolved. Thus, there are island species of rails and their close relatives that are little different from related species that can fly. For example, purple gallinules, which are relatives of rails, have colonized New Zealand at least three times from Australia. In the past, purple gallinules colonized the North Island, and, independently, the South Island. On both islands, new species evolved that were stout-legged and flightless, and heavier but otherwise very similar-looking to the ancestral purple gallinule. The North Island species is extinct, and the South Island species, now called takahē, survives in small numbers in alpine grasslands. More recently, purple gallinules have again colonized New Zealand from Australia, and in New Zealand they occupy different habitats from the rare takahē. These newest arrivals, called pūkeko in New Zealand, are not a distinct species. They have dispersed first throughout the North and South Islands of New Zealand and later to more remote islands in the Pacific, including Tonga. Ducks have also become flightless on colonizing islands. An endemic duck to

New Zealand, the brown teal, has, on different occasions, colonized two sub-Antarctic islands and their nearby islets. The local species that developed on Auckland Island and on Campbell Island have both become flightless but are otherwise almost indistinguishable from brown teal.

Flightless birds can become very large on islands. Examples are the extinct giant ducks, moa-nalo, of the islands of Hawai'i, which probably weighed 4–7.5 kg. A species found only on Kaua'i had a large, heavy bill like that of a tortoise. Three other species, which were found on one or more of the islands of O'ahu, Maui, Moloka'i, and Lāna'i, had serrated bills. Moa-nalo were grazing birds, utilizing habitats that, on continents, might be used by grazing mammals. Their ancestor, surprisingly, was a duck similar to the widespread Pacific black duck (a close relative of mallard) and this ancestral duck colonized Hawai'i quite recently, about 3.6 million years ago. That different species of moa-nalo were found on different islands implies that initially these birds could fly to colonize new islands and later became flightless. The development of flightlessness is probably why, on the Island of Hawai'i, there is no evidence of moa-nalo: they could not disperse to it. Instead, a goose related to the still extant nēnē colonized the Island of Hawai'i from North America and it too became a large and flightless grazing bird. None of the giant duck species survived after humans settled the islands of Hawai'i. The role played by these birds as grazers in determining the evolution of Hawaiian plants has now also been lost. Spines and thorns that are found on some Hawaiian plants, especially some species of the lobelias, *Cyanea*, could have been a response to the grazing pressure these birds exerted.

In New Zealand, the extinct moa birds were all flightless and some were very large. Females of the largest species probably weighed up to 250 kg and their backs were 2 m off the ground (their bone anatomy makes it seem unlikely that they held their heads high on their long necks). The females of the largest species were one-and-a half times taller and nearly three times heavier than the males; this size reversal is without parallel in any other land vertebrate. Not all of the approximately ten species of moa were this big; others were at most 1 m tall. They were browsers or grazers, and they occupied a variety of habitats from coastal dunes to forests to alpine grasslands on the large islands. One moa species had a sharp beak that could cut through twigs more than a centimeter thick and some trees and shrubs developed very high tensile strength in their branches, which may have been as a defense against moa. Throughout most of the 60 million years that moa were in New Zealand they were not subject to predation. However, about 2 million years ago an eagle colonized New Zealand, probably from Australia. From this ancestor a new species of eagle developed, called pouākai, which could attack and kill even the largest moa. Pouākai was the world's largest eagle, with a wingspan up to 3 m and weighing up to 15 kg. DNA from subfossil bones of pouākai reveals that it is most closely related to the smallest living eagle in Australia. Its evolution from an eagle one tenth its weight to the largest ever is likely to have been influenced by the absence of a competitor for its prey, moa, which may have been easy quarry. Neither moa nor their predator, pouākai, survived long after human settlement in New Zealand (see Chapter 5).

Whereas some animals on colonizing islands became giants, sometimes the converse trend occurred and animals became smaller. This is most apparent in mammals. The now-extinct Honshu wolf in Japan was much smaller than its Asian ancestor, the Eurasian wolf. Similarly, the fox species that is unique to the Channel Islands off the coast of California is smaller than the mainland species from which it derives. Some of the best examples are from extinct hippopotami and elephants found as fossils on some islands in the Mediterranean. Their ancestors walked to the islands during periods when the Mediterranean Sea was dry, then, when the islands were formed as sea levels rose; the isolated individuals became much smaller than their ancestors. This phenomenon, known as island dwarfism, results from the limited population size possible on small islands. If individuals are smaller, the island will support a larger number of them, all things being equal. Most species have a minimum number of breeding pairs needed to survive. In addition, the ability of the habitat to support a population fluctuates over time. Hence, for large animals, being smaller lowers the risk of extinction. In contrast, species that were already small are often freed from predation and competitors when they colonize islands. Becoming large reduces their ability to conceal themselves or fly, but on islands this matters less. On the mainland, other species occupy the larger size ranges: on an island, there is no barrier and thus resources for which they could not compete on the mainland become available. Up to the point of running the risk of having too small a population, small animals therefore become larger, longer-lived, and as a result, more conservative in their breeding. An example of this can be seen in the pūkeko example discussed above. The small, flighted, and fecund purple gallinule (or pūkeko) is the ancestor of the large, flightless, slow-breeding takahē.

Among plants, the tendency for herbs on continents is to have relatives that are woody on remote islands. We have already mentioned in this chapter the woody species of *Sonchus* and *Echium* of the Canary Islands: most other members of these genera are herbs. On islands world-wide, many members of the daisy family become woody. Woody daisies are not unique to islands, but they are often disproportionately species-rich on islands. For example, New Zealand and its outlying islands have 62 woody species in two genera (*Brachyglottis* and *Olearia*). The most likely explanation for this phenomenon is that on islands, especially temperate ones, there are few open or even disturbed habitats when compared with grasslands and wide floodplains on continents from which the herbaceous ancestors colonize. Thus, on colonizing islands, selective pressure operates on herbs to become perennial and tall in order to compete with the woody plants already present.

4.4.4 Strong connections between land and sea

Islands are surrounded by the sea and maritime influences can be seen in the kinds of plant and animal communities that are on islands. A particular feature of many small islands is the presence of seabirds, often in large numbers. These include penguins, cormorants, gulls and terns, gannets and boobies, puffins (Plate 4) and their relatives, albatrosses and shearwaters. Although a few species nest at the edges of continents, many seabird species nest exclusively on islands. For some, such as cormorants, that do not forage widely, the exclusive use of islands is driven by the proximity of food sources. For many others, islands, especially those without predators, are a safe breeding place because they nest on the ground. Until the chicks fledge, they are vul-nerable to predation while the adults forage for them. Few seabirds nest on the larger islands that we consider, such as the main islands of the British Isles, Japan, Jamaica, Puerto Rico, and New Zealand, but they are common on many offshore, smaller islands. Seabirds did nest over much of the main islands of New Zealand until human settle-ment (see Chapters 5 and 6).

Seabirds transport large amounts of nutrients to the land from the sea, as guano and spilled food. They can also cause large amounts of disturbance because many nest colonially, and on many islands these colonies consist of thousands or even tens of thousands of birds (see Chapter 3 and Box 4.2). Colonies of species such as penguins and cormo-rants that nest on the surface may accumulate large amounts of guano. Shearwaters and petrels (Fig. 4.13) nest in burrows below ground that can be more than a meter long and that are re-excavated frequently. When adults return to nesting colonies they trample and scrape the ground surface bare. Seals also cause high levels of disturbance on islands and likewise bring nutrients ashore. On sub-Antarctic islands, such as Campbell Island in the far south of New Zealand, Hooker's sea lions can disturb terrestrial habitats across the whole island.

Box 4.2. Noisy seabirds

By day Ruamāhuanui is a quiet place. This small island off the northeast coast of the North Island of New Zealand is a steep fragment of an extinct volcano that is now disintegrating. During the day, a few land birds like parakeets chatter in the tree crowns and cormorants and terns feed in the waters off the coast. But it was clear, even before I landed, that this is an island where seabirds breed: tens of thousands of them. The smell of ammonia wafted from the island. Guano from blue penguins streaked the boulders of the landing bay. Once I was ashore and among the shrubs or under the forest canopy, the evidence of seabirds was everywhere. It looked as though someone has taken a vacuum cleaner to the forest floor: the earth was bare (Fig. 4.13). Seabird burrows were everywhere so I had to tread lightly. I pirouetted ungainly from one tree trunk base to the next to avoid collapsing the burrows – not easy when there could be a burrow every square meter. Two hours after dark the cacophony of returning grey-faced petrels, northern diving petrels, and little shearwaters began. The birds had been feeding long distances out to sea and some of these birds were back in New Zealand waters for only the breeding season. The birds crashed through the forest canopy and sat stunned on the forest floor for a few minutes, then scuttled quickly the few meters to their burrows. Some of the seabirds blundered into my tent. Tuatara were also about. They dashed quickly from point to point or retreated cautiously, head first into a burrow. The crescendo of wailing from the seabirds arriving back or calling from their burrows went on through most of the night. There was no prospect of sleep for four of us staying on the island. Before dawn the adult seabirds had left. Some climbed trees, even with their webbed feet, to get airborne: telltale scratch marks were visible along some of the leaning tree limbs. As I left the island, the eggs and chicks were in the burrows and the island was still and quiet again. PJB

Fig. 4.13   Forests on Ruamāhuanui in the Ruamāhua (Aldermen)
Islands, northern New Zealand. The soils are full of burrows made by
grey-faced petrels (called oi by Māori); leaf litter has been scraped away
by the birds (see Box 4.2).

The high nutrient input around seabird colonies favors the
growth of certain plants (see Section 4.2.1). Some plants grow exclusively in the highly nutrient-rich, but often highly acidic, soils that
form around seabird colonies. A species of cress that is unique to New
Zealand is found around colonies of seabirds such as gannets. Botanists
who accompanied James Cook on his voyage to New Zealand in 1769
recognized cress as an edible plant and used it to ward off scurvy. This
cress is now rare on the large islands of New Zealand, possibly because,
compared with 1769, seabird colonies are much depleted. Cress is now
most common on small islands where seabird colonies are intact. The
maintenance of entire island plant communities can depend on seabird nutrient inputs. On the Aleutian Islands in Alaska, seabird communities promote growth of lush grassland. Arctic foxes are not native
to the Aleutian Islands but were liberated on some of them in the early

1800s to promote a fur trade. The foxes extirpated the seabirds which had been on these islands and when the high nutrient input from the seabirds ceased, the grassland was replaced by tundra that does not require high nutrient levels.

4.5 EXTINCTION ON ISLANDS

4.5.1 Extinction as an evolutionary process

Extinctions are as important as colonization and radiation of species as determinants of the evolution of the plants and animals that occur on islands. We have already discussed how tectonic processes can directly cause extinction. Species on continental fragments, such as New Zealand, underwent a major process of extinction when much of the land mass was submerged. Jamaica almost certainly lost all the plants and animals that it once had when the island was either totally or nearly submerged for several million years before it re-emerged. Similarly we have discussed how oceanic volcanic islands, once they have moved away from the hotspot, which gave rise to them, slowly weather, crumble, subside, and finally disappear below the sea, along with the endemic species that were confined to them.

Tectonic processes can cause extinctions more indirectly. Fossils from southern New Zealand from 16–19 million years ago, long after the land had separated from Gondwana, are from ecosystems unlike any in modern New Zealand. They suggest a landscape much more like modern Australia. Eucalypts, acacias, and casuarinas, which are characteristic of modern Australia, were common in New Zealand at that time. Fossil crocodiles and a small rat-sized primitive mammal have been found. All of these became extinct; no related species evolved from them, at least, none that survived to the present or even the recent past. What changed and why were these species lost from New Zealand? The landscape in which these species lived was probably a flat, weathered plain: windy, warm, and dry. Not long after the period in which these species were found, a major collision began between the Australian and Pacific plates. A process of mountain building began forming the Southern Alps from about 14 million years ago and shaping a landscape more like modern New Zealand (see Chapter 2). A result of mountain building was that the prevailing westerly winds were now intercepted by new mountain ranges. New Zealand became wetter and cooler as the global climate cooled and as the islands moved closer to their present latitudes. The new climatic conditions and the disruption of the old weathered plains by tectonic activity were inimical for the survival of many plants and animals, and the new conditions favored other groups of species.

Catastrophic events and natural disasters, such as large volcanic eruptions (see Chapter 3), cause local and sometimes complete extinctions. We have already noted that glaciations caused extinctions, as happened in Iceland. Extinctions can also occur for biological reasons. Some evolutionary processes that take place on islands result in species that are unfit and do not persist. For example, the large-antlered Irish elk became extinct 7000 years ago. Its demise was likely triggered by the effects of glaciations that made Ireland a more difficult environment for this deer. Its extinction could be because the cost of developing and maintaining antlers on males was too great, or because the reproductive condition of females gradually deteriorated. Finally, newly colonizing species, arriving over new land bridge connections, may have been competitively superior and driven the elk and other local species to extinction.

4.5.2 Modern extinctions

There are few islands in the world that humans have not visited. Islands that had evolved unique plants and animals over millions of years were impacted abruptly by contact with humans from about 20 000 years ago. This dispersal was brought about by improvements in boat building and navigation. Humans brought with them animals, such as various rat species, which were to have profound influences on island plants and animals even when humans did not settle permanently on the islands. Humans brought fire to islands that had little evolutionary history of this disturbance and, on islands that did, human fires were at unprecedented spatial scales. On islands big enough to support permanent human populations, resources such as water were diverted, and fertile soils were deforested and replaced with non-native plants as crops. Domestic animals and other feral animals were liberated and these included grazing mammals and efficient predators such as cats and dogs (Box 4.3). Some animals endemic to islands, especially large flightless birds, made for easy sources of protein and were hunted to extinction. The process is on-going, as we will discuss in subsequent chapters. The result is a loss of the unique plants (Fig. 4.14) and animals on islands worldwide that is without precedent in the last 65 million years.

Box 4.3. Only one plant left

The Three Kings Islands were named by the Dutch navigator, Abel Tasman, who sighted them on Advent in 1643. These islands rise steeply and are surrounded by strong currents that flow between the Pacific Ocean and the Tasman Sea. Manawa Tawhi, or Great Island, is the largest of these islands and when Abel Tasman sailed by it, he saw the Māori people who lived on the island standing along the cliffs, presumably rather perplexed by the sight of a sailing ship quite unlike the canoes that they knew. Māori people had lived on the islands since the early fifteenth century, clearing the original forests and cultivating the flatter ground there, and they remained on the island until about 1840. After Māori people ceased living on the island, the abandoned agricultural land began to return to forest. This process might have been swift but it received a major setback when the New Zealand government decided, in 1889, to liberate goats on the island so that castaways might have sustenance in case of shipwrecks. Indeed, there were terrible shipwrecks around these islands – the worst when the *Elingamite* went aground in 1902 in heavy fog, resulting in 45 deaths. The goats thrived and multiplied and biologists who visited the islands in the 1920s voiced concern about the damage the goats were doing to the island's unique plants and animals. In response, in 1946 the government sent riflemen to the island, who eradicated the goats. The damage the goats had caused, by leaving the forest understory bare or by cropping grasslands to a turf, had at least made it easier for people to get about the island, so when the botanist Geoff Baylis visited the island, he was able to explore much of it rapidly. In his reconnaissance of the island, in 1945, in a single day he discovered two plant species that were previously unknown – a sturdy vine, *Tecomanthe speciosa*, and a tree, *Pennantia baylisiana* (named after its discoverer). The tree has the dubious distinction of being mentioned in the *Guinness Book of World Records* as the world's rarest tree, because the tree that Geoff Baylis found is the only one, and it is a female. I visited the island in 2003 by helicopter, 57 years after the goats were eradicated. By 2003, the forest had grown and the understory was thick – getting about the island is much slower than it was in 1946. My botanist colleagues, Anthony Wright and Ewen Cameron, took me to see the *Pennantia*. It is a sturdy, healthy tree, about 5 m tall, with large bright green leaves (Fig. 4.14). It grows on a steep rocky slope above sea cliffs.

Box 4.3. (cont.)

What is its future? Although the tree is healthy, it cannot live forever and it does not appear to produce seed in the wild. A powerful cyclone could uproot this last surviving tree. Active human management on the island is difficult – it is expensive and difficult to get to the island. Cuttings have been taken from the tree and it grows in cultivation on the main islands of New Zealand. Some of these cutting-grown trees have produced male flowers, and some have set viable seeds. In early 2010, staff of the New Zealand government's conservation department took seed from the cultivated plants to the island and they hope these seeds will germinate and grow. PJB

Fig. 4.14   *Pennantia baylisiana* (the only wild tree left) with its discoverer, Geoff Baylis. Photo in 1982. The tree is located on Great Island, Three Kings Islands, New Zealand (see Box 4.3).

4.6 SUMMARY

The plants and animals that occur on islands differ in many respects from those found on continents and they command our attention because of some of their special attributes. Only a certain set of plants and animals can colonize islands naturally. Most plants colonize islands by their seeds being blown to islands, floating to them, or being transported by flying or swimming animals. Islands that were formerly parts of continents or close to them, such as the British Isles and Japan, have greater similarities between their plants and animals and nearby continents than distant oceanic islands do with the closest continent. Some plants and animals that colonize islands have reached them across land that was exposed when sea levels were low, such as occurred during ice ages. Then species were marooned on the islands created when sea levels rose.

Once populations of plants and animals are on an island, either by dispersal or by being marooned there, the island's environment presents new opportunities and challenges. The animals that graze plants on continents may be absent on islands and, as a result, there is no need for the types of defenses that plants often have against herbivores, such as spines. Over time, new populations of plants evolve that lack these defenses. Animals that colonize islands may lack predators and gradually lose their means of escaping those predators. For example, birds can lose their power of flight. Some of the plant and animal species on islands become giants compared with their ancestors, others become dwarfs. The distinctiveness of the kinds of plants and animals on islands is a function of the age of the island, its distance from continents or other islands, its area, and its topographical complexity. The nine island groups we consider span gradients of all of these features. Very remote islands with a range of topography, such as Hawai'i, have some extraordinary plants and animals. Distinctiveness also depends on what plants and animals reach an island: some species differ little across islands worldwide while others evolve to form new species rapidly. For those species capable of rapid evolution to take advantage of new habitats, single founding population of plants or animals can evolve into new species to take advantage of the range of microclimates provided by such topographically complex islands as Jamaica or Puerto Rico. In groups of islands such as the Canary Islands or Hawai'i, new populations can disperse to different islands within the group and diversify on each. In this way, dozens, even hundreds, of new species can derive from a single founding population.

Islands naturally erode, are disrupted by volcanic eruptions, and are periodically smothered in ice (see Chapter 3). These events cause natural extinctions, and determine the makeup of plant and animal communities on islands. Humans are, in evolutionary terms, a very recent arrival on most of the world's islands. The extinctions that humans have caused, by both direct and indirect means, and human introductions of other plants and animals are very important determinants of the plant and animal communities found on islands today.

SELECTED READING

Carlquist, S. (1965). *Island Life*. New York: Natural History Press.
Carlquist, S. (1974). *Island Biology*. New York: Columbia University Press.
Darwin, C. (1859). *On the Origin of Species*. London: John Murray.
Fukami, T., D.A. Wardle, P.J. Bellingham, *et al.* (2006). Above- and below-ground impacts of introduced predators in seabird dominated ecosystems. *Ecology Letters* **9**: 1299–1307.
Gibbs, G. (2006). *Ghosts of Gondwana*. Nelson: Craig Potton Publishing.
Hedges, S.B. (1989). An island radiation: allozyme evolution in Jamaican frogs of the genus *Eleutherodactylus* (Leptodactylidae). *Caribbean Journal of Science* **25**: 123–147.
Lack, D. (1976). *Island Biology, Illustrated by the Land Birds of Jamaica*. Oxford: Blackwell.
Losos, J.B. (2009). *Lizards in an Evolutionary Tree*. Berkeley, CA: University of California Press.
MacArthur, R.H. and E.O. Wilson (1967). *The Theory of Island Biogeography*. Princeton, NJ: Princeton University Press.
McGlone, M.S. (2006). Becoming New Zealanders: immigration and formation of the biota. In: R.B. Allen and W.G. Lee (eds.) *Biological invasions in New Zealand*, pp. 17–32. Heidelberg: Springer.
Millien-Parra, V. and J.-J. Jaeger (1999). Island biogeography of the Japanese terrestrial mammal assemblages: an example of a relict fauna. *Journal of Biogeography* **26**: 959–972.
Sadler, J.P. (1999). Biodiversity on oceanic islands: a palaeoecological perspective. *Journal of Biogeography* **26**: 75–87.
Tennyson, A.J.D. and P. Martinson (2006). *Extinct Birds of New Zealand*. Wellington: Te Papa Press.
Thornton, I. (2007). *Island Colonization*. Cambridge: Cambridge University Press.
Wallace, A. (1858). On the tendency of varieties to depart indefinitely from the original type. *Proceedings of the Linnaean Society* **3**: 53–62.
Whittaker, R.J. and J.M. Fernández-Palacios (2007). *Island Biogeography: Ecology, Evolution and Conservation*, 2nd edn. Oxford: Oxford University Press.
Woods, C.A. and F.E. Sergile (eds.). *Biogeography of the West Indies: Patterns and Perspectives*. Boca Raton, FL: CRC Press.
Worthy, T.H. and R.N. Holdaway (2002). *The Lost World of the Moa*. Christchurch: University of Canterbury Press.
Ziegler, A.C. (2002). *Hawaiian Natural History, Ecology, and Evolution*. Honolulu, HI: University of Hawai'i Press.

5

Human dispersal, colonization, and early environmental impacts

5.1 INTRODUCTION

Human settlement on the island groups relates closely to the islands' isolation. The least isolated island groups were colonized by humans earlier than remote islands. The timing of first human settlement on the island groups varies from hundreds of thousands of years ago (British Isles), to tens of thousands of years ago (Japan), to thousands of years ago (Tonga), to hundreds of years ago (New Zealand). The oldest dispersals were across land bridges to the British Isles and Japan, and the more recent were across seas by boats to all the other island groups. In this chapter, we discuss the island groups according to the chronological order in which they were settled, from the oldest to the most recent. We discuss when the islands were first settled and where the first settlers came from. We describe the landscape they encountered on arrival and where people first settled in this new landscape. We also outline what people did after they settled and what happened to the plants and animals on each of the island groups as a result.

5.2 WALKING TO BRITAIN

Startling recent discoveries from Pakefield, on the Suffolk coast in England, show that human ancestors (*Homo erectus*) were present in the British Isles 780 000 years ago. These discoveries date from a warm period between glacial advances and at the time the British Isles were connected by land to the rest of Europe. The evidence is in the form of worked flint artifacts embedded in deposits of ancient rivers. The deposits also include remains of animals present at that time, such as hippopotami, rhinoceroses, and elephants. These artifacts are 200 000 years older than the half-million-year-old fossils of *Homo heidelbergensis* found at Boxgrove in southern England. An accumulation of fossil evidence from the British Isles shows that various human ancestors were present during at least four warm periods between successive advances of glaciers during a period from 25 000 to 700 000 years ago. When glaciers advanced, most of the British Isles were covered in ice and human ancestors were either absent or left no trace. On each occasion when humans or their ancestors colonized the British Isles they did so by walking across land bridges from continental Europe (Fig. 5.1).

The most recent colonization of the British Isles by humans led to unbroken settlement until the present. This colonization began in the southern and central parts of England as the last ice age ended about 15 000 years ago. Understanding how humans interacted with the environment needs to be put in the context of major climate change that occurred after the end of glaciation and that caused changes in the landscape and its vegetation. The humans who walked to Britain at the end of the last ice age were Paleolithic hunter-gatherers and they encountered a tundra landscape inhabited by animals such as woolly mammoths, which were still present in northern England until 10 000 years ago. There is no evidence, however, that mammoths were a major focus of human hunting. Paleolithic people did hunt other animals of the tundra: bones of arctic hares, fashioned into awls, and the antlers of reindeer, a quintessential animal of tundra, are found in caves in central England and are 13 000 to 15 000 years old. Art painted by the early settlers on the ceilings of these caves that date from before 12 800 years ago depict red deer and other elongated, stylized figures.

Between 10 000 and 12 000 years ago, human settlement of the British Isles expanded slowly from the south and east to the north and west but only by about 100 km. The climate remained harsh, as the glaciers retreated and re-advanced. The vegetation during this period changed from tundra and shrubland to forest. The Paleolithic people of about 12 000 years ago in southern England had a diet that was high in animal protein and featured wild cow and red deer, but not wild horses or reindeer. Reindeer were hunted largely for their bones, which were used to make various tools and ornamental pieces.

After the final retreat of glaciers from the British Isles 10 000 years ago, sea levels rose rapidly and flooded much of the North Sea and the English Channel, but did not sever land links to continental Europe. Ireland was separated from the rest of the British Isles at this time. What had probably been a long lake in Ireland was breached to become the Irish Sea. By 9500 years ago, the glaciers had retreated

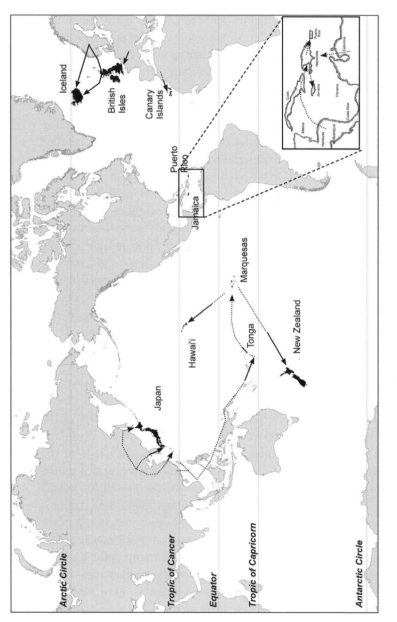

Fig. 5.1 Approximate migration routes of initial settlers to each of the nine island groups. Solid lines represent a high degree of certainty and dotted lines a lesser degree of certainty about the origins or migration routes of island settlers.

from the northern coasts of Scotland and birch forest covered most of the British Isles, except for the far north and west where tundra and shrubland still dominated. The seed sources came from forests that had survived beyond the maximum extent of the glaciers.

As this forested landscape developed, the hunter-gatherer Mesolithic cultures of the British Isles supplanted Paleolithic cultures from about 10 000 years ago. By about 8000 years ago, Mesolithic settlers reached some of the more remote parts of the British Isles by boat, such as Orkney and the island of Harris. A land bridge remained across the North Sea from the east of England to the Netherlands until about 6000 years ago, when it was finally breached by the rising ocean. By that time, almost all of Britain was forested, with mixed deciduous forests in the south, pine forests in the uplands of Scotland, and birch forests furthest north including the Outer Hebrides, Orkney, and Shetland.

Human impacts in the forested landscape were localized and clearances made by fire were on a small scale, often concentrated either near the coast or beside lakes. People relied on the sea or lakes for sources of protein, especially fish and waterfowl. The forests provided habitat for wild pigs and deer, as well as wild plants that were important sources of nuts, berries, and fruits, such as apples and pears. This abundance of food sources fueled the growth of human populations. Human-caused disturbances determined patterns of distribution and fluctuations in abundance of some dominant trees in the forested landscapes from between 5300 and 8500 years ago. For example, pollen records throughout the British Isles show increases in the abundance of trees such as alder and ash at local scales. These increases may be due, at least in part, to human disturbance because both of these trees are early colonizers of cleared forests. Conversely, it is improbable that the pine forests of Scotland (Fig. 5.2) or the hazel woodlands of eastern England were the results of large-scale human fires. The human population grew during the Mesolithic period and people made increasingly intensive use of food from forests, hunting deer, and gathering and storing hazel nuts. Food from lakes and the sea were also exploited, including shellfish in southwestern Ireland and oysters in Shetland. The Mesolithic period ended with a rapid transition to more sedentary life and the beginning of domestication of plants and animals, as we will discuss in Chapter 6.

Fig. 5.2   Scots pine forest near Inverness, Scotland. These forests were once widespread in Scotland but are now much reduced in extent.

5.3 JAPAN'S FIRST SETTLERS

Human ancestors reached Japan across land bridges from Korea (Fig. 5.1) and were present at least 130 000 years ago. As in the British Isles, successive glacial advances drove these human ancestors back to eastern Asia. Here we will focus on the most recent colonization of Japan by modern humans, which began the period of continuous settlement until today. The most recent human colonizers probably reached Japan around 37 000–38 000 years ago from the Baikal area of Siberia by crossing land bridges. This migration led to colonization of the islands of Honshu, Shikoku, and Kyushu. The settlement of nearby coastal areas of Russia occurred about the same time. To reach the other Japanese islands, such as Okinawa, at this time, people had to travel by boat. Indeed, the oldest human skeletal remains found in Japan are from caves in Okinawa, in the central Ryukyu archipelago to the south of Japan's main islands, with dates estimated at 32 000 years old. There is further evidence of early use of boats by these settlers; between 20 000 and 25 000 years ago, people journeyed from Honshu to Kozushima, an island 50 km off the coast, to bring back obsidian, a volcanic glass, for use as knives. The earliest evidence of human presence on Hokkaido, Japan's northern main island, is from at least 20 000 years ago; people may have crossed land bridges to reach Hokkaido.

The people who colonized Japan from 38 000 years ago were nomadic hunter-gatherers. Over time, the Paleolithic hunter-gatherer way of life developed to become the Jōmon culture of Japan, which began about 12 000 years ago and lasted until about 2000 years ago. People of the Jōmon culture were more sedentary than their Paleolithic ancestors and they made pottery; some, up to 12 000 years old, is among the oldest pottery in the world. The earliest Jōmon pottery vessels were likely to have been used over fires to boil and cook food. Vestiges of Jōmon culture persisted in Hokkaido until about 600–1000 years ago. The modern Japanese population on Hokkaido shows genetic traces of the Jōmon people. The Jōmon legacy is most prevalent in the Ainu people living today in northern Japan and the people of the central Ryukyu archipelago.

Human colonization in Japan, as in the British Isles, needs to be interpreted in the context of climate change. People first colonized Japan at the height of the glacial advances. The transition from Paleolithic to Jōmon culture occurred as glaciers retreated and any land or ice bridges ceased to exist as sea levels rose and the climate warmed. During the same period, there were substantial changes in animals and plants that occurred in Japan. For example, there were major changes in the Japanese mammal fauna after humans colonized. When humans first settled in Japan, the larger mammals present on these islands included mammoth, moose, brown bear, leopard, and a large deer. All of these and the steppe bison, aurochs, and another large deer became extinct about the time that Jōmon culture developed. At least three small mammals also became extinct during this period, including two voles and a shrew. Cave fossil deposits from lowland southern Honshu, which date from before human settlement, show that these extinct voles and other small mammals were present there when the climate was much colder. The small mammal species that still survive to the present day, such as the Shinto shrew, are now restricted to the coldest parts of Honshu in the highest mountains. It is therefore difficult to determine whether the extinctions of mammals that occurred after human settlement can be ascribed mainly to climate change, human activity as the Jōmon culture developed, or a combination of the two.

The vegetation of Japan also changed substantially before and during the 10 000 years of Jōmon culture. Before Jōmon culture developed, northern Hokkaido was largely covered by tundra and devoid of forest cover. From southern Hokkaido to north of Tokyo, and along the central mountains, a forest of spruce, beech, and elm species predominated, and southwestern Japan was covered in forests of deciduous oak, beech, and elm species. From about 4000 to 8000 years ago, the Japanese climate was warmer than at present and ocean levels rose. Warmer temperatures caused polar ice caps to melt and the sea flooded coastal lands and invaded river valleys, drowning forests. The forests of this warm period in southern Japan were comprised of mostly evergreen trees, especially evergreen oaks.

As the post-glacial climate of Japan became milder, natural resources became more abundant and the Jōmon culture thrived. The forested landscape changed from one of conifer dominance, when the landscape was cool, to one of deciduous trees as the climate warmed. Humans developed a more sedentary lifestyle indicated by an apparent shift from reliance on natural rock shelters in the earliest Jōmon settlements to the development of small communities of pit dwellings. From about 3000 to 5000 years ago, a period of climatic stability in Japan enabled population growth. During this period, human populations were concentrated near the coasts and were heavily dependent on seafood, especially shellfish, but they also hunted tuna, mackerel, seals, whales, and dolphins. The people were also adept bow hunters of deer, boars, and bears in the forests, using dogs to assist in their hunts.

What impact did the Jōmon people have on the forests of Japan? At local scales, these people created temporary clearings where pines and other secondary vegetation, especially ferns, grew in the midst of forests dominated by other species, such as oaks. These clearings were created around dwelling sites or encampments for gathering food. For example, people made clearings in coastal forests close to sites from which they collected shellfish. Once the clearings were abandoned, vegetation developed to become like that of the surrounding mature forests.

Over time, Jōmon cultures were able to manipulate the natural environment in which they lived to favor certain native plants that could sustain their populations. This was the case in the cool climates of southern Hokkaido and northern Honshu where people could store nuts of wild deciduous trees, including acorns, chestnuts, and walnuts. From early in the Jōmon period, people also mastered the ability to render Japanese horse chestnuts palatable, through a complex process to remove toxins. The Jōmon people favored such tree species and maintained them, even as the climate cooled in later periods. In part because of the abundance of food, high-density populations of people developed with complex spiritual rituals and a social hierarchy. Agriculture also developed with the cultivation of grains, beans, and gourds, as seen in evidence from 9500 years ago from a site in central Honshu.

About 6000 years ago, Jōmon settlers sailed south from Kyushu, and colonized some of the smaller, more remote Japanese islands, such as the island of Tanegashima in the northern Ryukyu archipelago. As Jōmon culture developed, the Jōmon people apparently moved across the sea, probably prompted by reliance on seafood. There is evidence for this movement in similar styles of pottery across southern Kyushu, the Ryukyu Islands, and Korea. A thriving trade in shells from the Ryukyu Islands developed in exchange for bronze implements from the main islands of Japan. There is also evidence of trade between the Ryukyu Islands and the Yellow River region of China between 27 000 and 4000 years ago. There is little known about the impact of people on these island environments, although bones of dogs, deer, and boar are found in settlement sites.

In the southern Ryukyu Islands, Okinawa was colonized by humans from up to 32 000 years ago, but this period of settlement apparently ended about 19 000 years ago. After that date there is no evidence of human settlement again until 10 000 to 12 000 years ago. People of the Jōmon culture island-hopped from the main islands of Japan only as far south as Okinawa and several nearby smaller islands. The 230 km stretch of open water between Okinawa and Miyako-jima far exceeded distances between islands to the north. The tropical Black Current, flowing north, and the greater frequency of typhoons at this latitude, perhaps presented additional barriers to colonization. Human settlement patterns in the Ryukyu Islands contrast with the patterns of dispersal of plants and animals in the islands. The relatively narrow Tokara Strait in the northern Ryukyu Islands is a sharp boundary between plants and animals that colonized either from the north or from the south to the Strait and not across it (see Chapter 4), but the Strait posed no barrier to human dispersal. In contrast, the straits south of Okinawa that were a barrier to human dispersal, because the seas were too dangerous to cross by boat, were not a major barrier to dispersal for plants and animals. Jōmon people also reached the Kuril Islands, north of Hokkaido at least 7000 years ago. While the Ryukyu and Kuril Islands were colonized in prehistoric times, the more remote volcanic islands of the Ogasawara Islands remained uninhabited until the nineteenth century. These islands lie 1000 km south of Tokyo and extend nearly to the Tropic of Cancer.

5.4 SETTLING PUERTO RICO AND JAMAICA

The Greater Antilles in the West Indies include the islands of Cuba, Hispaniola, Jamaica, and Puerto Rico. Cuba and Hispaniola were settled around 6000 years ago by people who probably traveled by boat from the Yucatán Peninsula in Central America, or from South America (Fig. 5.1). These people were foragers and fishermen. A second group of people migrated from South America into the Lesser Antilles and reached Puerto Rico about 3800 years ago. Evidence of different origins of these first waves of settlers to the Greater Antilles comes from different skull dimensions. These data suggest that Cuba may have been colonized separately from the other islands in the Greater Antilles. Jamaica was not settled during this period. Together, these early settlers are considered the Archaic people.

Population densities of the Archaic people were low and they left few remains in the caves in which they sheltered. They subsisted on seafood, including fish, turtles, and shellfish from reefs and habitats close inshore, and on seasonally available foods from the forests. They began making pottery about 4000 years ago and lived in permanent villages. There is evidence of increasing fire frequency in the forests after human settlement from about 3000 years ago. The Archaic cultures cultivated a variety of tree crops, including avocado, and root crops, including an edible cycad. They also introduced some useful tree species, such as canistel, from Central America.

There was apparently another group of settlers that reached Puerto Rico about 2500 years ago and that came from northern South America. Their route may have been directly to Puerto Rico and the Virgin Islands, rather than island-hopping the smaller islands of the Lesser Antilles. These colonizers are referred to as the Arawak or Saladoid people and their cultures are defined by distinctive new styles of pottery. Like the Archaic cultures, they also settled largely along the coast and relied heavily on marine resources, but they also brought with them horticultural crops and animals from South America. These people were the first to colonize Jamaica, where the oldest archaeological sites are 1300 years old, much more recent than those of Puerto Rico, Hispaniola, or Cuba. It is unclear why Jamaica was one of the last islands of the Caribbean to be settled, but it may be that rough sea conditions around Jamaica discouraged settlement until the period when large canoes were constructed on the large neighboring islands. The Taíno were the most recent dominant culture of the Greater Antilles, including Puerto Rico and Jamaica, from about 800 years ago (Fig. 5.3). Their origins are debated: they may derive directly from the Arawak people or from a further wave of colonization from South America.

Fig. 5.3   Caguana ceremonial ball court of the Taíno Indians in Puerto Rico. In addition to its recreational uses, the courts, built between AD 1200 and 1500, were used for ceremonial dances, religious rites, and perhaps astronomical observations.

The first people to arrive in Puerto Rico and Jamaica encountered wholly forested islands. Each of the islands in the Greater Antilles had unique groups of animals that lived in the forests, notably giant rodents and, on some, giant sloths. All of these giant animals became extinct after humans arrived on the islands. These extinctions may not have been rapid: recent evidence suggests that a now-extinct giant sloth on Cuba probably survived for up to 1000 years after human settlement. On Jamaica, which was settled by people much more recently, not only the giant rodents, but also a species of monkey, a flightless ibis, and a large bird of prey became extinct after human settlement. Some marine resources, such as manatees, were locally extirpated.

Human populations moved to the forested interiors of these islands only fairly recently. Settlement of the islands' interior may have been prompted by a depletion of marine resources but it may also have been triggered by a period of climatic instability, which made life at the coast more difficult. Settlement of the inland foothills on Puerto Rico began about 1600 years ago. The limestone hills and plateaus in Jamaica were settled about 1100 years ago. Agricultural systems strongly influenced by South American practices of local deforestation and plantations reached the Caribbean around

2400 years ago: a practice known as slash-and-burn agriculture. This required felling trees using stone axes, burning the fallen trees, and then sowing crops in the ash bed. This kind of agriculture required a population to move to new land every 10 to 12 years, after the soil has been exhausted of available nutrients for growth of crops. New and more intensive agricultural practices were developed by about 1800 years ago, including the use of fertilized mounds, terracing, and irrigation. After the Spanish conquest ended Taíno agriculture in Jamaica, forests with a distinctive tree composition developed on these sites.

The intensive agriculture depended on crops from South or Central America brought to the islands by the settlers. Chief among the crops was cassava, but settlers also brought maize, tobacco, peanuts, and papaya. They brought animals including dogs, opossums, and agouti. Although agriculture required the clearance of forests, the forests were still important sources of food and in Jamaica there is evidence that some tree species that provided food were cultivated. For example, an upland species of guava was planted in the lowlands. The forests were also important as sources of timber, especially for construction of canoes, which enabled deep-sea fishing and trading among the islands. The canoes were large: in Jamaica, a traveler with Columbus measured one that was 29 m long and 2.5 m wide. The timber of other trees indigenous to the islands, such as lignum vitae, was used to construct stools used in religious rituals.

5.5. THE FIRST CANARY ISLANDERS

People first colonized the Canary Islands between 2000 and 3000 years ago and they came by boat from North Africa, probably from Western Sahara (Fig. 5.1). Recent genetic studies confirm that these people, the Guanches, shared the same ancestors as the Berbers, who now live in the mountainous regions of Morocco and Algeria. Later contact with people from northwest and sub-Saharan Africa had a limited genetic influence on the Guanche population.

A single main settlement may have resulted in all the islands being colonized from east to west, in a stepping stone pattern. An exception is that the Guanches of one of the most western islands, El Hierro, had genetic affinities closest to those of the easternmost islands, Fuerteventura and Lanzarote. By the time of European contact and colonization of the islands in the fourteenth century, each island had a distinct culture and language. Similarly, the archaeological remains of the Guanches differ across the whole archipelago, with those on adjacent islands showing the greatest similarities.

The Guanches' ancestors brought with them in their boats plants and animals not native to the Canary Islands. Grain crops such as wheat and barley were introduced. Taro may have been introduced at this time (Plate 9). Animals that were introduced included livestock, goats, sheep, pigs, cats, dogs, and mice. Although trees on the islands could have been used to construct canoes or boats, there is little evidence that these people returned to North Africa. Likewise, journeys between the islands were probably infrequent because of the dangers in navigating the oceans around the Canary Islands, where there are strong currents and constant trade winds. Such conditions discouraged fishing from boats, so the Guanches relied on coastal fishing and gathering shellfish.

The Guanches practiced Neolithic agriculture on the islands. Their life was based mainly on shepherding of livestock (principally goats), cultivation of grains, gathering seafood from the coast, hunting small animals, and collecting wild fruits and plants. The islands' forests were important for fuel wood, and charcoal records show that the Guanches used a wide range of the islands' trees for fuel. They had no access to metal ores and their tools were of stone. Obsidian was an important resource for knives and there were at least two major obsidian quarries on the island of Tenerife. The life expectancy of Guanches was not long: for example, on La Gomera, 60% of people died before the age of 35. Diet varied among the islands with greater dependency on seafood on small islands such as El Hierro compared with larger islands such as Gran Canaria. On Gran Canaria, the population was far more dependent on agriculture and supported a much larger, denser population than on the small islands. There were an estimated 50 000 people (or 30 per km^2) at the time of Spanish conquest. Dependence on agriculture came at a cost because rainfall is low and irregular and the islands were subject to plagues of locusts originating in nearby Africa. Many of the Guanche skeletons found on Gran Canaria show signs of malnutrition.

Until recently, little was known about the vegetation history of the Canary Islands and the effects of Guanche settlement on the vegetation of the islands. A record of pollen from Tenerife shows that until the time of human settlement there was a forest of oaks and hornbeams in the same valley as the modern city of La Laguna: such a forest no longer exists on the islands and oaks and hornbeams are no longer found in the Canary Islands. It is highly likely that the

Guanches burned these forests about 2000 years ago to make way for pastures and for agriculture, with hornbeam becoming extinct within 400 years and oak persisting until the Spanish conquest. In place of these native oak and hornbeam forests, the native laurel forests expanded. The impact of Guanche settlement, with livestock, fire, and introduced mammalian predators was also apparent in the effects on the native animals of the island. A giant rat and a giant lizard up to 1 m long on Tenerife, and the lava mouse on Fuerteventura became extinct soon after the Guanches colonized the islands, as did some land birds: a finch on La Palma, a quail on four of the islands, and the long-legged bunting on Tenerife. The bunting was apparently flightless; this is unusual because it was a passerine (or songbird). Only two other flightless passerines are known to have existed, both species of New Zealand wrens, and they too became extinct after human settlement. Two seabirds – species of shearwaters unique to the Canary Islands – also became extinct; one of these, the lava shearwater of Fuerteventura, persisted for probably 1000 years after settlement.

5.6 EARLY POLYNESIA – THE SETTLEMENT OF TONGA

The oldest dates of human settlement in Tonga are between 2790 and 2820 years ago. These settlements were from Tongatapu, the largest island of the archipelago; this is also the western end of modern Polynesia. The sites on Tongatapu, and others, slightly younger, on the Ha'apai islands, contain artifacts of the first settlers, known as Lapita who may have been cultural ancestors of modern Polynesians. The Lapita people derive their name from a distinctive style of pottery first developed 3200–3450 years ago far to the west of Tonga, in the Bismarck Archipelago along the northern coast of Papua New Guinea. As they traveled east, the Lapita people settled islands in modern day Vanuatu, New Caledonia, and Fiji, leaving behind their pottery as evidence of their settlement. The oldest fragments of Lapita pottery on Tongatapu are virtually indistinguishable from pottery of the same age from the Santa Cruz Islands, at the northern end of Vanuatu, so it seems likely that this was the point from which they left in canoes to colonize Tonga.

The Lapita people share the same ancestors as the modern indigenous communities of the coast and mountains of the island of Taiwan. According to one theory, the ancestors of the Lapita people began their southward migration from Taiwan (or near it) toward New Guinea about 5200 years ago, with pulses of movement then pauses during settlement (Fig. 5.1). The Lapita pottery styles that they refined later derive from styles developed in coastal Southeast Asia. In the Bismarck Archipelago, and during their migration east through Vanuatu, New Caledonia, and Fiji, they coexisted with long-established Melanesian civilizations on most of those islands. On reaching Tonga, for the first time they encountered islands where humans had never settled before. The colonization of Tonga and other nearby island groups, such as Samoa, led to human coloniza-tions of all of Polynesia: islands within a vast area of the Pacific Ocean that extends from Hawai'i in the north to Easter Island (Rapanui) in the east to New Zealand in the south.

The Lapita people settled in coastal sites in Tonga in small ham-lets of no more than three or four family groups, where they ate birds, sea turtles, fish, and shellfish. They took advantage of a previously unexploited ecosystem and have been described as "opportunistic foragers." Their experience of settling islands where these resources were unexploited no doubt was a powerful incentive to explore and exploit the resources of distant islands and it was in this way that the 800 km length of the Tongan archipelago was settled within one or two centuries.

One of the hallmarks of the Lapita people, and of the Polynesian navigators that they became, was that they transported not only them-selves by canoe to colonize new islands, but that they brought a large stock of crops and animals when they arrived to settle each new island. These included at least 50 plant species, including two important root crops, yam and taro, which derive from Southeast Asian ancestors. This self-sufficiency was necessary because the islands of the tropical Pacific Ocean have few edible plants. In the earliest stages of settle-ment of Tonga by the Lapita people it appears that agricultural activ-ities were localized and the people used slash-and-burn techniques.

The ancestral Polynesians also took animals on their voyages of colonization – pigs, chickens, dogs, and rats – the first three for food, the last as either emergency food or as "stowaways." Modern genetic techniques have been applied to populations of pigs, chickens, and rats found from Southeast Asia and Pacific islands in order to shed further light on the origins of the people who introduced them to the islands and the routes they took. Recent studies show that the pigs transported by the Polynesians and by the Lapita people before them, derive from wild ancestors in Southeast Asia, possibly Vietnam. Studies of variation in the DNA of the Pacific rat (Fig. 5.4), which has been widely introduced throughout the Pacific, support a view of a

Fig. 5.4   Pacific rat in a Hawaiian forest.

slow migration of modern Polynesians' ancestors through Southeast Asia to the Bismarck Archipelago and east through Melanesia, followed by rather rapid colonization of the rest of the Pacific. Seeds gnawed by rats and preserved in wetlands can also be used to date the arrival of people on islands because it is highly unlikely that Pacific rats, which are poor swimmers, reached the islands by other means than on voyaging canoes. During the earliest stages of colonization of Tonga, there is evidence of chickens at the oldest archaeological sites but not of pigs or dogs, which may have been introduced later.

In Tonga, the developing Polynesian culture maintained links with Melanesian Fiji to the west: for example, there is evidence of exchanges of pottery during the early period of settlement of Tonga. However, the Lapita pottery style ceased to be made in Tonga about 2650 years ago, and pottery making altogether was abandoned in Tonga and throughout Polynesia, before the settlement of Hawai'i and New Zealand. The reasons for the abandonment of pottery are unknown: perhaps cooking methods changed from boiling food to baking it. Population growth in Tonga between 1550 and 2650 years ago led to the settlement of the interior of the islands, the growth of agriculture, and increasing reliance on imported domesticated animals (chickens and pigs) over wild birds and other terrestrial animals as sources of food.

The effect of human colonization by Lapita people on the animals that were unique to Tonga was catastrophic. The palaeontologist David Steadman has described the effect as a "blitzkrieg" because of the suddenness and high numbers of extinctions that took place. For example, of 34 land bird species in the Ha'apai islands in Tonga present at the time of human colonization, at least 21 did not survive beyond the Lapita period. On the large island of 'Eua, there were 27 forest-dwelling bird species before human colonization: now there are only nine species. Other animals besides birds became extinct as a result of human colonization, such as fruit bats and lizards, including a large iguana. The species that became extinct were those unique (endemic) to individual islands, large-bodied species, and those bird species that lived disproportionately on or near the ground, especially those that were flightless.

Extinctions of animals in Tonga that resulted from human colonization have been well documented but less is known about the effects of human settlement on the vegetation of the islands. On the northern islands of Vava'u in Tonga, the amount of charcoal found in lake sediments increased immediately after human settlement and this was probably caused by people burning the forests. Pollen from the same site shows that, in response, some tree species that profit from disturbance, and species such as coconut, which were useful crops to the settlers, became more abundant. At the same time, native species that were formerly abundant declined rapidly and are now very rare in the modern forests.

As human populations increased, so did the pressure to deforest the islands to make way for most of the staple crops on which Polynesian culture depended (Plate 10). Polynesian agriculture was an intentionally "transported landscape," which depended on introduced plants and animals and was replicated on islands throughout Tonga, and throughout Polynesia. Some Tongan islands and others elsewhere in Polynesia were deforested rapidly by fire before populations became very large; on some small, dry islands, a single fire set by the first person to arrive could have resulted in their deforestation. While direct deforestation was instrumental in the decline of forest trees, recent research has drawn attention to the indirect consequences of human settlement on forest composition. For example, the extinction of large-bodied pigeons and fruit bats probably resulted in a lack of dispersal of fruits of certain forest trees. The extinction of nectar-feeding birds could have resulted in some native plants no longer being pollinated as effectively and therefore not reproducing. The introduced Pacific rat feeds selectively on the seeds or seedlings of certain tree species, such as palms, and one result of this behavior could be failure of these species to regenerate. Finally, the loss of seabirds, either directly from hunting or as a result of predation by rats, could have led to reduced inputs of marine nutrients to the land that may have been important in maintaining the pre-human vegetation (see Chapter 4). Although some species became extinct or less abundant, species that were absent before human settlement, such as the Pacific imperial-pigeon, have flown to Tonga and become established after human settlement. Possible reasons for this pattern of colonization could include the extinction of species previously present with which the new colonists had been unable to compete, modification of the landscape by people that provided more suitable habitat for the new colonists, or just the unpredictable nature of long-distance dispersal – the new colonists just never happened to land in Tonga.

As Tongan society developed, it began to rely on many native plants for fiber, medicine, natural oils, and construction materials. Seafood (Plate 11) and wild game, especially pigeons, were important food sources for Tongans. Therefore, the Tongans designated areas of forest or populations of animals for permanent conservation; other resources were protected for only certain lengths of time. For example, an area might be protected long enough to allow regrowth of trees or to allow recovery of fish populations after exploitation. The Tongan society developed a strong social hierarchy and laws to enforce these rules. Rules to ensure permanent or temporary protection of natural resources were likely to have been developed in response to awareness of how some of the more vulnerable plants and animals had either been driven to extinction or severely reduced in numbers.

5.7 REACHING THE EDGES OF POLYNESIA – THE DISCOVERY AND SETTLEMENT OF HAWAIʻI

Polynesians probably reached the islands of Hawai'i about 1200 years ago, but the timing of arrival is still the subject of research. After the colonization of Tonga and Samoa, Polynesians colonized eastern Polynesia. The ancestors of the Hawaiians probably sailed in double-hulled canoes from the southern Marquesas Islands in eastern Polynesia to Hawaiʻi (Fig. 5.1). The Polynesians who colonized the islands of Hawaiʻi brought with them in their canoes a cargo of their traditional agricultural crops, and animals. By the time they colonized the islands of Hawaiʻi, Polynesians also had new crops to bring with them: sweet potato (Fig. 5.5) and bottle gourd. These crops are native to South America. Polynesian voyagers may have reached South America and brought these crop plants back with them to eastern Polynesia.

Fig. 5.5   Sweet potato garden on 'Eua, Tonga.

As the Polynesians migrated throughout the islands of the tropical Pacific, the "transported landscape" approach to agriculture became increasingly refined. The islands of Hawai'i offered ideal conditions for Polynesian agriculture and the scale at which it could be practiced because the islands were large and offered gradients in climates and rainfall, fertile volcanic soils, and locally abundant water. Initial settlements, as they had been in Tonga, began at the coast and proceeded inland into the valleys. The Hawaiians went on to develop highly sophisticated food production, including large-scale, irrigated taro cultivation; large-scale, organized cultivations of sweet potato; and aquaculture using artificial fishponds (Fig. 5.6).

Soon after colonization, forest burning began, as shown in charcoal and pollen evidence from O'ahu and Kaua'i. Recent analysis of fossil evidence shows that forests, at least on a limestone plain in the rain shadow of O'ahu, disappeared soon after human settlement. The introduction of Pacific rats is likely to have been a factor in the forests' disappearance because the rats ate the seeds of some tree species. Trees that seem to be particularly vulnerable to predation of seeds by rats include *Pritchardia* palms (Fig. 5.7). Nearly half of the 23 *Pritchardia* palms endemic to the islands of Hawai'i are now critically endangered, with some confined to tiny, rat-free islets.

The Hawaiian population grew exponentially during the period from 500 to 800 years ago and may have reached about 300 000 people. In this period, deforestation continued in the lowlands and began on the lower slopes of the volcanoes, transforming the native vegetation into a mosaic of managed landscapes. By about 400 years

Fig. 5.6   The ancient He'eia fishpond on the western shore of O'ahu, Hawai'i. Note mangroves in foreground and partial restoration and mangrove removal in background.

Fig. 5.7   A rare Hawaiian *Pritchardia* palm growing on a cliff overlooking the western coast of O'ahu, Hawai'i.

ago, at least 80% of all the lands in the islands of Hawai'i below about 450 m in elevation had been extensively altered by people. The soils of the high rainfall areas of the main islands were subject to nutrient loss through leaching and the low rainfall areas were subject to wind erosion: areas of intermediate rainfall were the zones most favored for agriculture. These areas could support very high human population densities. For example, in the Halawa Valley of Moloka'i, densities have been estimated at up to 250 people per km^2 – similar to those supported by rice farming in tropical Asia. These populations and their impacts were highly concentrated. Only 10% of the total area of the islands of Hawai'i was intensively used by humans before European contact. For example, there is little sign of direct impacts on the higher altitude forests. Deforestation by the Polynesian settlers of Hawai'i, coupled with seed predation by Pacific rats, resulted in the near-complete destruction of the tropical dry forests on the leeward (southern and southwestern) sides of the main islands; only vestiges of dry forests remained in places such as on Lāna'i by the time of European contact. Some plant species probably went extinct as a result of deforestation. For example, one type of pollen that is present in the fossil record before human settlement was unknown in the modern plants of Hawai'i. Then in 1992, only two wild plants of a previously unknown shrub in the legume family were discovered on a tiny steep islet off the coast of the island of Kaho'olawe. Pollen from this new species closely matched fossil pollen from the lowlands of O'ahu, Maui, and Kaua'i; on those islands the pollen may be of related but now-extinct species.

Human settlement precipitated a wave of extinctions in the land animals of the islands of Hawai'i, following the same "blitzkrieg" pattern as in Tonga. In addition, the introduction of predatory rats in particular, and of pigs and dogs, also caused extinctions well beyond the immediate zones of human settlement. As in Tonga, the large, flightless bird species were among the first extinctions. Soon after settlement, humans hunted to extinction the giant, flightless moa-nalo ducks, of which there were unique species on most of the larger islands. Hunting also caused the extinction of a flightless ibis and a flightless goose on Maui. Of 74 species of land birds present at the time of human settlement across the islands of Kaua'i, O'ahu, and Moloka'i, only 33 survived until the time of European contact. Fossil evidence from caves and sinkholes on the south coast of Kaua'i are indicative of the extent of the loss of species. Until human settlement, there were more than 40 indigenous bird species. Of these, nearly half are now extinct and only a handful of the remaining species are now found in the immediate vicinity: most of the remainder survive only in the mountains. In the same caves, there are the remains of fourteen endemic land snails; nearly all of these are presumed to be extinct. Modification of the lowlands by the Hawaiians transformed areas into fire-prone grasslands, for example the coastal plain around the modern city of Honolulu. These new, open habitats were novel and could be colonized by new species. Short-eared owls colonized the islands of Hawai'i only about 1000 years ago. This owl is likely to have reached Hawai'i before human settlement, but the forested landscape of the islands would have been unsuitable habitat for it. New grassland habitat created after human settlement, and suitable prey introduced by people, such as the Pacific rat, meant that the owl could then colonize the islands successfully.

5.8 SETTLING THE LARGEST ISLANDS OF POLYNESIA – REACHING AOTEAROA (NEW ZEALAND)

So far in this chapter we have taken a chronological approach to the sequence of settlement of islands by people but we depart from that here. Among the islands we consider, New Zealand (Aotearoa) was probably settled more recently than Iceland, but we shall continue the thread of Polynesian voyaging and discovery (Fig. 5.1). Until the discovery of New Zealand, Polynesian settlement had focused on tropical or subtropical islands. Not only was New Zealand by far the largest area that they settled (more than 16 times the size of the islands of Hawai'i, for example) but its temperate climate offered new challenges for settlement.

The most likely date for Polynesians' arrival and settlement in New Zealand is 730 years ago, making it the last inhabitable, large land area settled by humans anywhere in the world. The evidence for the date of settlement derives from bones of the Pacific rat or from seeds that the rats gnawed. This date coincides closely with the dates of the earliest human activities as revealed from mounds of food waste (middens), and of human-lit fires and is consistent with the time required to produce the apparent variation in the genetic structure of the Polynesian descendents, the modern Māori. However, the date of settlement is contentious: some researchers contend that rats were introduced 1200 years earlier, and that this fits a model of a long period of discovery by Polynesians, followed by exploration, initial settlement, and colonization.

The Polynesians who voyaged to New Zealand came from eastern Polynesia and they traveled in double-hulled canoes. Their exact point or points of origin in eastern Polynesia are unknown. As in their other intentional voyages, they brought their cargo of crops and animals including Pacific rats and dogs. If they brought pigs and chickens on their voyage to New Zealand, the animals either did not survive the journey or did not last long after arrival, because there is no trace of either pigs or chickens in archaeological sites. As on smaller tropical Pacific Islands, the earliest archaeological sites in New Zealand are along the coasts. A fish lure made from mother-of-pearl at one of these sites (dated to around 700 years ago) must have come from tropical Polynesia. For the people who settled at the coast, the marine resources, hitherto unexploited, were an important part of the diet in the developing culture of Māori in New Zealand.

The eastern parts of New Zealand, in the rain shadow of the large mountain ranges, were forested until settlement by Māori. These forests generally had been unaffected by fires before human settlement, because lightning strikes were rare. The first Māori settlers set fires in the rain shadow zone and this led to the near-elimination of fire-sensitive forests and shrublands from all but the wettest districts of the eastern parts of the North and South Islands. The rate of deforestation was rapid: most forests were eliminated within 100 years of settlement. Nearly one-third of the South Island's area, or about 50 000 km^2, was deforested and its vegetation replaced by native grasses and shrubs that could survive fire. The rainforests of northern and western New Zealand were less easily cleared. Most of the growing Māori population lived in these warmer and wetter regions, where they practiced agriculture. Clearance of some coastal areas of rainforest for development of agriculture took place up to 300 years after initial colonization.

The "transported landscape" that had been a successful part of Polynesian settlement on tropical islands was not easy to apply in New Zealand. Gardening of root crops sustained many Māori communities in northern New Zealand and followed the model of Polynesian slash-and-burn agriculture in Tonga and Hawai'i. However, some of the crops that were likely to have arrived with the voyagers, such as breadfruit and coconuts, could not grow in New Zealand's temperate climate. Others, such as the bottle gourd and taro, could grow only in the northern parts of the country, especially near the coast. Cultivation of sweet potato required careful management as far south as Banks Peninsula on the east coast of the South Island. So while New Zealand had a greater extent of land than Polynesian cultures had ever known, its capacity to support traditional Polynesian agriculture was limited. Sources of starch would be barriers to colonizing the interior of both main islands and the southern South Island. Māori in these areas relied on the roots of the native bracken fern as a source of starch, perpetuating its management on the best soils as a source of food, and elsewhere practicing fire management to encourage bracken growth along routes that gave access to the interior.

The abundance and size of native wild birds in New Zealand were unprecedented for the new Polynesian colonizers and contrasted with the difficulties encountered in practicing agriculture. The largest flightless moa weighed up to 250 kg and must have been easy to catch and kill. We surmise this from the large numbers of moa slaughtered and butchered in sites throughout the North and South Island. However, this abundance of easily available meat on land was short-lived: there are almost no bones of moa found that are younger than 600 years old and it seems likely that Māori drove all ten species of moa to extinction within about 100 years of settlement.

The "blitzkrieg" for native wildlife was not confined just to moa. Other birds may have become extinct as a result of hunting and the giant eagle, pouākai, presumably died out when moa, its main prey, was slaughtered and its forested habitats were cleared. Many more species of animal were eradicated because of the introduction of the Pacific rat. Large, flightless insects that roamed the forest floor became extinct: we know about these animals because of remains encased in pumice and buried under volcanic ash from before human settlement. Many reptiles, including tuatara and lizards, were eradicated from the main islands of New Zealand, probably because of rat predation. Small birds, including seabirds, and small land wrens, some possibly flightless, became extinct soon after human settlement. By the time of European settlement, 27 species of birds had disappeared, even in the most remote parts of the main islands of New Zealand.

Māori continued seafaring and established trade routes among their various scattered tribes for stones such as jade, obsidian, and argillite, as well as for food. Māori settled many of the islands close to the New Zealand coast: in the north of the country some of these islands were ideal for raising crops brought originally from the tropics because frosts were rare and because the soils had been fertilized by seabirds. To the south, Māori traveled 1000 km to the Auckland Islands. They had most likely already discovered the Kermadec

Islands, 1000 km north of the North Island, before settling New Zealand, but then returned to the Kermadec Islands from New Zealand. Evidence of these travels comes from the presence of obsidian on these remote islands. Obsidian was quarried in only a few places in New Zealand and was an important stone used for knives. On their voyage to the Auckland Islands, as far south as Polynesians have been known to travel in their era of voyaging, Māori and their dogs survived on the islands for at least one season, living on seals and seabirds. Māori also colonized and settled the Chatham Islands, 800 km east of the South Island and a new civilization (the Moriori) developed there. Extinction of species of animals and plants unique to the Chatham Islands resulted from human colonization of these islands.

5.9 COLONIZING ICELAND – THE NORSE OUTPOST

The first people arrived in Iceland 1160 years ago and permanent settlements began soon after. The ancestors of the modern Icelanders were Norwegians, but it is also likely that their route to Iceland was through Norse colonies in the western Scottish islands (Fig. 5.1). Irish slaves were also part of the founding populations of Iceland so the modern Icelandic population has Scandinavian and Celtic origins. Although literary references and place names suggest that Irish monks had sailed to the island about 150 years earlier than the Norse and could have settled, there is no archaeological evidence for this. Iceland became a staging point for expansion of the Norwegian empire to the west, to Greenland and at least as far as the Labrador coast by 1000 years ago. During this period of expansion, new settlers fleeing conflict in Norway arrived in Iceland. Initial settlement of Iceland was along the coast but within two generations, families were settling in the interior.

The Icelandic settlers brought a cargo of domestic and wild animals with them, including sheep, cattle, pigs, goats, horses, dogs, cats, and probably mice. The modern Icelandic sheep (Fig. 5.8) has scarcely altered from the stock that the Norse brought, a breed that was common in Scandinavia and northern Britain at the time of settlement of Iceland. Likewise, the modern Icelandic horses (Fig. 5.9) have changed little from those brought initially. In contrast to many other island settlers, the first Icelanders did not introduce rats to the island – Norway rats arrived only within the last 200 years. It is intriguing that there were outbreaks of bubonic plague in Iceland during the Middle Ages, which

Fig. 5.8   Icelandic sheep are cold-hardy and bred for meat, wool, and milk. They are among the purest breed of sheep in the world.

Fig. 5.9   Icelandic horse. Iceland does not allow import of other breeds of horses or the return of Icelandic horses taken abroad, keeping this breed pure and mostly disease free. Long-lived and hardy, they have adapted to the sometimes harsh Icelandic climate.

must have taken place in the absence of rats as hosts for fleas. The first Icelandic settlers also brought barley and perhaps flax. They also accidentally introduced non-native weeds, insects, and earthworms. Like the Polynesians, the Norse settlers applied a model of a transplanted agricultural system in their settlement of islands in the North Atlantic, including Shetland, Orkney, the Faroe Islands, and Iceland.

The first settlers of Iceland encountered an island that was about 25% forested. The forests of birch, rowan, and willows were in the lowlands (below 400 m), and heaths and shrublands, as well as local grassy areas, occurred in the uplands. Volcanic landscapes, glaciers, and high-altitude areas that could not support vegetation made up about one-third of the island. There were very few native land animals – arctic foxes were present and polar bears were rare visitors, arriving on ice floes, and there were no native grazing animals. At the coast there were sea mammals, including walruses, large seabird colonies, and migratory nesting birds.

The settlers deforested the lowlands to make way for pasture for livestock and to cultivate barley. In the first centuries after colonization, a political order was established in Iceland with a parliament, strong chiefly authority, and laws. Chiefly authority was closely tied with numbers of cattle and grazing land that a chief owned. In the earliest stages of settlement, the Norse colonizers relied more on cattle than on sheep and goats for food. Pig farming persisted for only a little more than a century after settlement. An immediate consequence of deforestation was that the removal of forests resulted in short-stature regrowth that was readily browsed, and, without tall trees, the timber available for construction or boat-building was soon depleted. The forests were important as sources of firewood for cooking and to enable people and their livestock to survive the winter cold. Forests also were felled to make charcoal, which was needed for the working of iron. Recognition that a resource of timber needed to be conserved may be a reason for slow rates of deforestation apparent in some pollen records and it may also be a reason for the reduction in numbers of pigs and goats during the first centuries after settlement. Nonetheless, the forests were maintained and managed for at most 450 years, and now less than 3% of the land is forested.

The application of Norse agriculture to the type of volcanic soils found in Iceland had not been attempted elsewhere. The effects of deforestation and grazing, coupled with exposing soils wind and water, initiated chronic problems of soil erosion that have become steadily worse. The initial effects of soil erosion began soon after settlement and were localized, but became widespread by about 300 years ago. As a result, an estimated 40% of the soil present at the time of settlement has disappeared. Former Norse farms are now ruins amidst large gravel beds (Fig. 5.10).

Fig. 5.10   Severe soil erosion on Icelandic farms caused by centuries of over-grazing.

Barley cultivation in southern Iceland ended about 600 years ago and this is probably due to the "Little Ice Age" of 500 to 860 years ago and also to soil degradation. In other Norse settlements, additions of animal manure maintained the productivity of the soil, but perhaps in Iceland either livestock numbers or human labor was insufficient to maintain the practice. Only haymaking ensured the survival of live-stock, and sheep became the dominant grazing animal. Other natural resources faced mixed fortunes after settlement. There are records from northern Iceland of harvests of wild waterfowl and their eggs, as well as seals, fish, and beached whales. Walrus remains were common in archaeological sites of the early Norse settlers but walruses no longer breed in Iceland and may have been exterminated by the settlers.

In contrast to other islands, the settlement of Iceland resulted in few if any global extinctions because the plants and animals of Iceland were all recent colonists after the retreat of ice sheets 10 000 years ago and because most are widespread in the land areas of the North Atlantic (see Chapter 4). However, one infamous extinction occurred quite recently in Iceland. The great auk was a large seabird, once wide-spread in the North Atlantic Ocean, but easily preyed upon because it was flightless. By 500 years ago, its numbers had been severely depleted for sources of down. By the 1830s, the last small breeding colony was in Iceland, but by this time their pelts and eggs were valued by museums throughout Europe. The birds in this last surviving colony were all killed for museum specimens: the last adults were strangled and their eggs smashed on the islet of Eldey in 1844.

The early settlement of Iceland had made use of the natural capital of soils, forests, and other natural resources and led to the rapid development of an independent, strong society. However, the consequences of deforestation, unsustainable grazing, soil degradation and erosion, unstable climatic change, bubonic plague, and disasters related to the island's many active volcanoes all took their toll. The population was further subjugated by rule from Scandinavia and by 250 years ago it had become an impoverished rural province with a population of about 30 000.

5.10 SUMMARY

There are consistent patterns in how the first people to arrive on islands settled and modified their environment across all the island groups that we describe. Whether people arrived across temporary land bridges or by boat, the first settlements began at the coast and moved inland over time. In all cases, from boreal Iceland to tropical Jamaica, the first settlers generally encountered forested landscapes, although in the British Isles and in Japan, the first settlers arrived as forests were developing after the retreat of glaciers. Forests were important in providing firewood for the early settlers, food sources, and habitat for food and game. However, for islands settled by people who were already familiar with agriculture, the forested landscape was an impediment to food production and was cleared.

The initial stages of settlement in forested islands in some cases were sustained for thousands of years (for example, in the British Isles and Japan), and many hundreds of years in Puerto Rico, with rather subtle changes in forest vegetation. More recent settlements in the Pacific islands and in Iceland have resulted in rapid changes in forest cover, and in some cases almost complete removal of some forests, within the space of a hundred years or less.

In general, the more isolated the islands and the longer their isolation, the more devastating was the arrival of people for the plants and animals of the islands. Recent human settlers also brought increasingly sophisticated agricultural tools and were more efficient at clearing land. In long-isolated New Zealand, where much of the vegetation was combustible but had very rarely had natural fires, forests over large areas almost completely disappeared in less than a hundred years. In Jamaica and Puerto Rico, where unique land mammals had evolved in isolation from nearby Central America, very few survived long after the arrival of people. The unique, large land birds of Tonga, New Zealand, and Hawai'i all became extinct rapidly after human settlement, as did large lizards on Tonga and the Canary Islands. When settlement of islands also involved the transport of introduced animals, both desired (livestock) and stowaways (rats), this caused or hastened the demise of native plants and animals, from trees that may have failed to regenerate because of the destruction of seeds by rats in Hawai'i, to the stripping of plant cover followed by erosion in Iceland. The consequences of human colonization of islands have been universally damaging to natural environments. Iceland and New Zealand were settled only about 430 years apart, yet despite their vastly different cultures, differences in tools (metal for the Icelanders, stone for Māori) and boats, as well as differences in the cargo of introduced plants and animals that they brought, the result in terms of deforestation and transformation of the native plants and animals was remarkably similar. Only very remote and small islands still remain relatively unaffected by humans. In the next chapter, we will address the continued environmental degradation from the development of agriculture, extensive deforestation, and global trade. Some remarkably parallel stories emerge, despite the huge differences among the nine island groups in timing of human arrival, climate, types of resource extraction, and choice of agricultural crops.

SELECTED READING

Anderson, A. (2009). The rat and the octopus: initial human colonization and the prehistoric introduction of domestic animals to Remote Oceania. *Biological Invasions* **11**: 1503–1519.

Atkinson, L.-G. (ed.) (2006). *The Earliest Inhabitants: The Dynamics of the Jamaican Taíno*. Mona: University of the West Indies Press.

Burley, D.V. (1998). Tongan archaeology and the Tongan past, 2850–150 B.P. *Journal of World Prehistory* **12**: 337–392.

Campbell, I.C. (1992) *Island Kingdom: Tonga, Ancient and Modern*. Christchurch: Canterbury University Press.

Cuddihy, L.W. and C.P. Stone (1990) *Alteration of Native Hawaiian Vegetation: Effects of Humans, their Activities and Introductions*. Honolulu, HI: University of Hawai'i Press.

Fitzpatrick, S.M. (ed.) (2004). *Voyages of Discovery:The Archaeology of Islands*. Santa Barbara, CA: Greenwood Publishing.

Fitzpatrick S.M. and A. Anderson (2008). Islands of isolation: archaeology and the power of aquatic perimeters. *Journal of Island and Coastal Archaeology* **3**: 4–16.

Fitzpatrick, S.M. and W.F. Keegan (2007). Human impacts and adaptations in the Caribbean Islands: an historical ecology approach. *Earth and Environmental Science Transactions of the Royal Society of Edinburgh* **98**: 29–45.

Habu, J. (2004). *Ancient Jomon of Japan*. Cambridge: Cambridge University Press.

Kerr, G.H. (2000). *Okinawa: The History of an Island People*, rev. edn. Tokyo: Tuttle Publishing.

Kirch, P.V. (2000). *On the Road of the Winds: an Archaeological History of the Pacific Islands before European Contact*. Berkeley, CA: University of California Press.

Kirch, P.V. and J.G. Kahn (2007). Advances in Polynesian prehistory: a review and assessment of the past decade (1993–2004). *Journal of Archaeological Research* **15**: 191–238.

McGovern, T.H., O. Vésteinsson, A. Friðriksson, A., *et al.* (2007). Landscapes of settlement in northern Iceland: historical ecology of human impact and climate fluctuation on the millennial scale. *American Anthropologist* **109**: 27–51.

Martin, P.S. and D.W. Steadman (1999) Prehistoric extinctions on islands and continents. In: MacPhee, R.D.E. (ed.) *Extinctions in Near Time: Causes, Contexts, and Consequences*, pp. 17–56. New York: Kluwer.

Milberg, P. and T. Tyrberg (1993). Naïve birds and noble savages: a review of man-caused prehistoric extinctions of island birds. *Ecography* **16**: 229–250.

Pearson, R. (1977). Paleoenvironment and human settlement in Japan and Korea. *Science* **197**: 1239–1246.

Rouse, I. (1992). *The Tainos, Rise and Decline of the People Who Greeted Columbus*. New Haven: Yale University Press.

Smith, C. (1992). *Late Stone Age Hunters of the British Isles*. Abingdon: Routledge.

Smith, K.P. (1995). Landnám: the settlement of Iceland in archaeological and historical perspective. *World Archaeology* **26**: 319–347.

Steadman, D.W. (2006). *Extinction and Biogeography of Tropical Pacific Birds*. Chicago, IL: University of Chicago Press.

Stringer, C. (2006). *Homo Britannicus: The Incredible Story of Human Life in Britain*. London: Penguin Group.

Tennyson, A.J.D. and P. Martinson (2006). *Extinct Birds of New Zealand*. Wellington: Te Papa Press.

Totman, C. (2000). *A History of Japan*. London: Blackwell.

Vitousek P.M., T.N. Ladefoged, P.V. Kirch, *et al.* (2004). Soils, agriculture, and society in precontact Hawai'i. *Science* **304**: 1665–1669.

Ziegler, A.C. (2002). *Hawaiian Natural History, Ecology and Evolution*. Honolulu, HI: University of Hawai'i Press.

6

Intensifying human impacts on islands

6.1 INTRODUCTION

In this chapter, we follow the development of human civilizations across the nine island groups until modern times. We address the changing relationship of civilizations on the islands with their environments. In Chapter 5, we described the first civilizations that developed on the islands after their initial settlement, but, for many of the islands, new waves of settlers arrived later. Often there were violent clashes between the residents and the later settlers and major changes in island environments were brought about by new settlers. It is beyond the scope of our book to record in detail the human histories of each of the island groups: entire libraries are devoted to the histories of the British Isles and Japan alone. Our intention is to highlight common trends across the island groups.

On settlement of an island, people first required that the finite resources of an island were sufficient to meet the basic human needs of food and shelter. If these needs could be met, civilizations developed. Populations have grown on some of the islands that we consider, sometimes to quite extraordinary densities, but there are consequences of such growth for the island environments. Agricultural societies developed from the hunter-gatherer civilizations that preceded them in the British Isles and the main islands of Japan, and this may also have been the case in Puerto Rico. For the other island groups we consider – Canary Islands, Jamaica, Tonga, Hawai'i, New Zealand, and Iceland – the first settlers arrived with agricultural practices already developed and introduced them to the islands, as we have seen in Chapter 5.

In the British Isles and in Japan, the transition from the hunter-gatherer to agricultural civilizations also brought about profound changes to the environments of these islands. Yet there are contrasts between the ways in which agricultural landscapes developed in the British Isles and in Japan, imposed in part because of their different geographies. Japan is much more mountainous and topographically complex than the British Isles (see Chapter 2); therefore, proportionately less land area is amenable to agriculture in Japan. We begin this chapter with an analysis of the intensification of human impacts on the British Isles and Japan, linking these impacts to developing human cultures. Then we describe how the Canary Islanders turned from self-sufficiency to cash crops that changed their environmental impacts. Next, we discuss how the environments of the Caribbean islands of Puerto Rico and Jamaica were impacted by Spanish conquerors, slavery, and cash crops. Important impacts on the Polynesian islands of Tonga, New Zealand, and Hawai'i included resource harvesting by and for Europeans and the advent of European agriculture. Finally, the accumulated environmental damage to Iceland by its original colonists poses big challenges to modern day Icelanders who still struggle with the effects of deforestation and soil erosion.

6.2 DEFORESTATION OF THE BRITISH ISLES AND THEIR CONVERSION TO AGRICULTURE

The origins of agriculture in the British Isles lie in the Neolithic Revolution that started in the Middle East and gradually spread across Europe. This revolution began about 9000 years ago when, as the historian Alfred Crosby put it, people "reached out, grasped, and manipulated whole divisions of the biota around them." The change in culture was marked by people bringing wild plants into cultivation and domesticating wild animals. The Neolithic Revolution of the Middle East involved the cultivation of cereal grasses, and these were introduced to the British Isles as its Neolithic culture started between 5000 and 6000 years ago. For example, cultivars of barley, whose wild ancestors are native to the Middle East, were introduced to the British Isles about 5000 years ago. To achieve the changes of the Neolithic Revolution required a large degree of social organization. For example, the introductions of crops and domestic animals, including sheep and goats, entailed Neolithic farmers bringing them by boat from continental Europe across the English Channel, the North Sea, and the Irish Sea.

The shift from hunter-gatherer to Neolithic culture in the British Isles was a decisive break from total reliance on the wild plants and animals native to the islands to a principal reliance on introduced plants and animals, although native species remained an important part of the diet and culture of Neolithic people. As we have seen in Chapter 5, the hunter-gatherer cultures of Britain and Ireland learned to clear the forest cover of the islands at local scales to favor habitat for certain wild foods and game. Neolithic agriculture, with its newly introduced cereal crops, used the same process of small-scale forest clearance; the crops were grown and harvested and when soils were depleted the clearings were abandoned and forest cover gradually returned. Neolithic agriculture did not reduce the total forest cover of the British Isles substantially, but it did change the composition of forests. The introduction of sheep and goats favored the regeneration of tree species with seedlings that were resistant to grazing, and did not favor tree species that were unable to survive or reproduce with grazing, forest clearance by fire, and extended periods of cultivation. These are the likely reasons for the decline of elms in lowland forests from about 5700 years ago, evident in fossil records of pollen. Conversely, some native herb species that had previously been confined to harsh tundra environments became widespread and abundant as a result of Neolithic agriculture, while new herbs native to southern Europe invaded and naturalized with the introduced cereal crops. Neolithic farmers reached many of the remote parts of the British Isles, including the Shetland and Orkney Islands (Fig. 6.1). On the Shetland Islands, people were present from about 4500 years ago,

Fig. 6.1   Ring of Brodgar, a Neolithic stone circle in the Orkney Islands that was built about 4000 years ago.

and settled and farmed in a largely forested landscape. On the main island of Shetland, between about 3600 and 4000 years ago, settlers had reduced the extent of the birch and hazel forest, but when they abandoned agriculture, the forest returned.

Some areas cleared during the Neolithic period in the British Isles were to become its first areas of permanent lowland grasslands and pasture, and these were maintained by grazing animals. For example, evidence from pollen and land snails indicates that small parts of the landscape in southern England had already been transformed from forest to shrubs and open pasture by about 4300 years ago; for example the area around Stonehenge, before and after its construction.

The development of metalwork was a technological change that led to significant change in people's capacity to modify the landscape. The beginning of the Bronze Age in the British Isles, from about 4100 years ago, is likely to have resulted from the arrival of immigrants from continental Europe, who brought bronze-making technology to the islands. The copper and tin ores needed to make bronze were found in the British Isles, although from even the beginning of the Bronze Age some ores were imported from continental Europe. Trees needed to be felled to manufacture charcoal that was required to melt the component metals and make the bronze alloys. Bronze implements included flat axes, which enabled people to fell forests more efficiently than stone implements had allowed, and farm implements (such as plows), which made tending and growing crops more efficient. The advent of metal culture in the British Isles enabled expansion of agriculture, which gathered in pace and extent over the next 1500 years, and the transformation of the British Isles from a forested landscape to the largely pastoral one that prevails to this day.

The developing Bronze Age culture became more organized and hierarchical over time (Fig. 6.2), and its agricultural basis moved from subsistence to one where fields were designated for particular purposes and forms of land allotment developed. Domestic animals, which had been solely a source of meat in Neolithic times, were harnessed to plow fields. The animals were used as sources of wool and milk, and their manure was used to maintain the fertility of fields. The new society, increasingly geared to agricultural production, embarked on wholesale clearance of the forested landscape.

During the early to middle part of the Bronze Age, from about 3800 years ago, the climate of the British Isles was mild and this enabled the settlement of the uplands of the islands. Forests were cleared

Fig. 6.2   Bronze Age funerary monument (Ballowall Barrow, Cornwall, British Isles) built to provide a tomb for a leader of the community. Stone walls were added by the excavators in the 1870s to highlight features of the original mound.

from most of the uplands of England, Wales, and Ireland and walled fields created for agriculture. However, clearing these forests exposed the thin soils beneath them and this had severe environmental consequences. Widespread erosion resulted, so that 1000 years after forest clearance and farming began, much of the topsoil of upland plateaus had eroded into valleys and floodplains. The soils that did not erode became more weathered and, over time, no longer supported pasture but became acidic. The vegetation changed to dwarf shrubs such as heather and bilberry, the moorlands that characterize these landscapes today. Even after agriculture was abandoned, the forests never returned. In extreme cases, soils were lost completely. Clearance of the forests of the limestone plateau of the Burren in western Ireland during the Bronze Age exposed soils that were used for agriculture until they eroded away through natural fissures in the limestone. The result of Bronze Age clearance of the forests, as the palaeoecologist Neil Roberts put it, is that "all that is left is bare limestone pavement, incongruously criss-crossed by Bronze Age fields with no soil inside them!"

    Widespread deforestation of the lowlands during the Bronze Age began earlier and was even more thorough than occurred in the uplands. In contrast to clearance of the uplands, deforestation of the

lowlands led to agriculture that was sustainable and highly productive. As iron tools replaced bronze tools during the Iron Age, from
about 2800 years ago, the rate of deforestation increased. As a result,
the forest cover of the lowlands of the British Isles was almost totally
removed during the later part of the Bronze Age and the Iron Age that
followed, between 2000 and 3000 years ago. During this period, the
Celtic cultures of the British Isles flourished. Further introductions of
crops, such as varieties of wheat, resulted in even greater productivity
of this agricultural landscape and this led to a large increase in human
population. For example, the fertile soils of the floodplains around
modern London were farmed and densely populated during the Iron
Age. The population of the British Isles increased from about 14 000 at
the beginning of the Bronze Age to about 2 million people by AD 43,
when the Romans invaded and subjugated most of the Celtic people.

Deforestation caused by the explosion in agriculture during the
Bronze and Iron Ages, and the burgeoning population that it fueled,
resulted in extinctions of forest-dwelling animals. The aurochs, the
wild ancestor of modern domestic cattle, is unlikely to have survived
in the British Isles beyond the Bronze Age; aurochs survived much
longer (until 1627) in continental Europe. Brown bears, which were
rare in the British Isles compared with continental Europe, probably
became extinct some time before the Romans invaded. Conversely,
mice were introduced during this period; their first certain remains
from the islands date from the Iron Age.

The agricultural landscape and patterns of land use established
in the Bronze and Iron Ages endured throughout the period of Roman
occupation, its subsequent collapse, and the period of feudalism that
followed. Romans continued the process of deforestation in the more
remote parts of their Empire. For example, along the border with
Scotland, there is evidence from pollen records of deforestation in
the vicinity of Hadrian's Wall, and from the use of tall timbers in the
construction of the wall (Fig. 6.3). After the collapse of the Roman
Empire, there were localized areas of forest regeneration. The Romans
had brought many plants to England, nearly all of which died out or
were maintained only in gardens. An exception was chestnut, which
is native to southern Europe but became fully naturalized in England
after its introduction by the Romans. It continued to invade relatively remote woodlands long after the Roman Empire collapsed. Ship
rats (Fig. 6.4), which are native to southern India, had spread to the
Mediterranean by about 2200 years ago and were introduced by the
Romans to the British Isles.

Fig. 6.3   Hadrian's Wall, west of Housesteads, Northumberland, England.

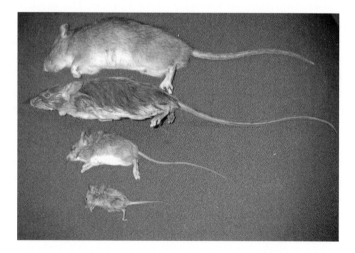

Fig. 6.4   Four common rodents introduced to many of the island groups by humans. From the top: Norway rat, ship rat, Pacific rat, house mouse.

Shortly after the Norman invasion of England, the conquerors undertook a comprehensive census in 1086, the Domesday survey, which recorded the deforested landscape that they encountered in England and Wales. The Normans noted that only 15% of the landscape was wooded. The feudal society that the Normans instituted was superimposed upon the land-use patterns set in the Iron Age.

The Norman kings and feudal barons, and those that followed during mediaeval times, took control of the remaining forests of England and Wales. Forests then ceased to be a resource held in common by the people as they had been during the Iron Age. The kings and the aristocracy controlled many of the forested or woodland areas as hunting domains for deer and other game and the Normans introduced new game species, such as fallow deer, originally native to the Middle East, and pheasants, originally from central Asia. Hunting domains in no way guaranteed sensible use of the game animals: King Henry III's profligate slaughter for a Christmas banquet in 1251 included 630 deer, 115 cranes, and 200 wild boar. A few years later he ordered the last remaining wild boar in England to be killed for a friend. Cranes, already a luxury food item in this period, ceased breeding in England soon after. The Normans exterminated native wolves in England and Wales by about 1300. Royal and feudal control of forests resulted in an increase in monetary value of timber and firewood, so that practices of woodland conservation and management to favor desirable species were instituted; forests were protected by draconian laws to prevent their degradation. Nevertheless, by 1350 the forested area of England and Wales had declined further to about 10% of the landscape.

The Normans also introduced rabbits, which were originally from southern Europe, to the British Isles. The earliest rabbits introduced were small and incapable of digging their own burrows. They were nurtured for meat, often on islands off the English and Welsh coasts. However, soon after rabbits were introduced, they began to damage tree seedlings by browsing them, and, over time, they became important determinants of the plant composition of grassland and heathland communities. The greatest ecological influence of rabbits occurred in the nineteenth and twentieth centuries.

Feudal agriculture and the growth of the monasteries during the medieval period prevented re-growth of forests. Selective grazing by sheep in upland areas drove the composition of the upland grassland and moorland toward a small number of plant species that could resist grazing, and that dominate many areas to the present day. As human populations grew from about 4 million to about 6 million between 1500 and 1600, so did the need for more agricultural land. English plantation settlers in Ireland rapidly cleared most of the island's remaining forests, which are estimated to have occupied about one-eighth of its landscape in the early 1600s and only one-fiftieth of its area 200 years later. Agriculture did not inevitably expand. For example, in northwestern England in the Forest of Bowland, expansion of agriculture was followed by a collapse of human populations during the fourteenth and seventeenth centuries, which was due to exhaustion of resources and outbreaks of plague. Some re-growth of forests, comprised of alder and birch, occurred when agriculture collapsed. Mostly, however, as populations grew, so did pressure on the islands' natural resources.

Population pressures, the growth of capital-intensive agriculture, and improved technology all resulted in the wholesale conversion of natural salt marshes and wetlands to agriculture. The destruction of most of the wetlands that comprised the Fenland of East Anglia, in eastern England, is an extreme example. Dutch engineers were employed by the Earl of Bedford to drain the Fenland and the process, begun in 1637, reduced the Fenland area from about 3380 km² to about 980 km² within 200 years. By the beginning of the twentieth century, only about 100 km² remained. The peaty soils, once drained, were well-suited to agriculture, but have shrunk so much as a result of drainage that now substantial areas of these soils lie below sea level, and have to be protected from inundation by levees. Maintenance of the small, isolated remnants of the original Fenland as wetlands is a challenge because they are now surrounded by drained areas at lower elevations. In other parts of the British Isles, the use of clay pipes to drain wetlands from 1800 onward greatly expanded the areas available for agriculture.

Agricultural reform that began in the late 1700s led to creation of enclosed fields throughout parts of lowland England where there had been extensive open fields (Fig. 6.5 and Plate 12). This change resulted in increasing concentration of land ownership, which in turn led to depopulation of parts of rural England, and migration to cities or emigration. This was also a period of rapid technological changes that further boosted agricultural production that, in turn, supported the burgeoning population of the islands (Fig. 6.6). In Scotland, lowlands and their land management were affected by similar agricultural reform laws. Feudal agriculture was the main mode of land use in the upland parts of Scotland and its outer islands, in which tenants practiced pastoral farming at small scales. Widespread expulsion of tenant farmers took place during the 1700s and continued until the 1870s. The concentration of land ownership resulted in depopulation and subsequent management of much of upland Scotland as exclusive hunting domains.

Land-use changes caused modern extinctions of wolves in Scotland (1621) and Ireland (1720), either from direct hunting or from loss of their main prey (deer) as a result of deforestation. Other

Fig. 6.5   Limestone farm walls, near Malham, Yorkshire, England.

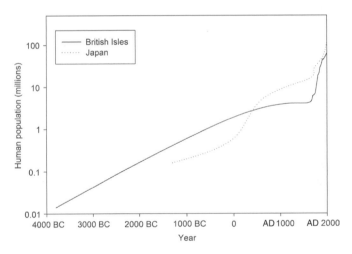

Fig. 6.6   Population growth in the British Isles and Japan.

species have become rare in the British Isles. For example, polecats, martens, and wildcats have all declined in abundance since the mid-1800s. In the last 300 years, 15 species of flowering plants and ferns have become extinct in the British Isles as well as about 40 lichen species. Lichens were particularly prone to local extinction caused by air pollution from industrialization and coal burning. At the same time, new species were added as a result of deliberate and accidental introductions. For example, sycamore, a maple tree from central Europe, was introduced as a decorative tree in the 1600s and is now a major component of many young forests in the British Isles. The grey squirrel, native to North America, was introduced deliberately in 1876 and now occurs widely throughout the British Isles, where it has largely replaced the native red squirrel. One consequence of the introduction of the grey squirrel is consumption of hazel nut crops, and the resultant regeneration failure of the hazel tree, which had been a major component of the islands' native forests (see Chapter 5). Muntjac deer were introduced from China in 1922 and are now common and widespread in lowland England. Norway rats, which despite their common name, are originally native to China, are a relatively modern and accidental introduction. Norway rats reached England by ship in the early 1700s (Fig. 6.4). They have thrived in cities where they have displaced ship rats to the countryside. The growing cities and industrial landscapes of the British Isles have provided new habitats, often favored by newly introduced plants that have escaped from cultivation, such as the butterfly bush (buddleja), which has spread along the gravel beds along railways, and giant hogweed, which spread initially along canals.

In the nineteenth and twentieth centuries, the urbanization of the British Isles continued, spurred by the Industrial Revolution and fueled by the societal upheaval caused by rural land reform and wars. Between 1801 and 1901, the population of the British Isles increased from 16 million to 42 million (Fig. 6.6). Historic uses of natural resources, including techniques of woodland management, declined, and deer populations subsequently increased because of low hunting pressure. Agriculture transformed the landscapes of the British Isles and is still a major land use, but today most of the population now has little contact with agriculture and people are heavily reliant on imported food (see Chapter 7).

6.3 JAPAN: A CIVILIZATION FOUNDED ON RICE CULTIVATION AND FOREST MANAGEMENT

Japan's hunter-gatherer societies, like those of the British Isles, made the transition to agriculture by modifying their techniques of localized felling of forest and by fire management (see Chapter 5). In part, the development of agriculture in the late days of the Jōmon culture was achieved by domesticating wild grains that were native to Japan, and growing a variety of crops of fruits, seeds, leaves, roots, or stems: at one site the plant species that were cultivated between 2500 and 3000 years ago included gourds, beans, hemp, and peaches. In

Kyushu, people began to grow grains (buckwheat and barley) that may have been brought from continental Asia. Climatic cooling during this period probably led to a more sedentary life as wild foods may have been less reliably available. However, the techniques of domesticating wild plants and cultivating them were then practiced more widely as the climate warmed from about 2500 years ago.

The transition to sedentary agricultural life was revolutionized by the introduction of hardy varieties of short-grain rice, probably brought from the Korean Peninsula. This rice required wet fields for cultivation, and the first evidence of its cultivation in Japan is from about 2400 years ago, close to the modern city of Fukuoka in Kyushu. The introduction of wet-rice cultivation was a defining characteristic of the Yayoi culture, which replaced the Jōmon culture. The culture spread, taking rice cultivation northward along both coasts of the island of Honshu and slowly inland; this migration stopped in central Honshu, north of which was too cold for reliable cultivation of rice. Initially, wet-rice cultivation was practiced alongside dry-field cultivation of older crops, and the more sedentary lifestyle that resulted from rice cultivation enabled people to keep domesticated animals, such as pigs. Cultivation tasks in the early stages of Yayoi culture were mainly the province of children and the elderly, while men continued to hunt wild game and gather wild foods. Diverse agriculture and the supplements of wild food fueled population growth in Japan from an estimated 160 000 people 3300 years ago to between 600 000 and 1 million 2000 years ago.

The burgeoning population (Fig. 6.6) placed greater demands on land. Naturally wet areas suitable for rice cultivation were soon exhausted and, as a result, construction began of dykes and irrigation dams to bring water to elevated fields. By at least AD 300, rice production had improved because of the development of the technique of transplanting seedlings into orderly rows in wet fields, which made it easier to keep the fields free of weeds. The large human populations in some areas resulted in the depletion of sources of wild foods close to habitation. Greater effort was needed to obtain wild foods, so their importance in diets diminished. During the development of Yayoi culture, iron tools imported from continental Asia or fashioned in Japan from imported iron ore, improved the efficiency of farming. Iron axes accelerated the rate at which land could be deforested. Smelting metal required firewood. The lowlands and foothills of Japan were thus transformed. The plains of southern Japan, previously covered in mixed forests that included trees in the laurel family and alders, were

Fig. 6.7   Planting rice near Yakushiji Temple, Nara, Japan.

deforested and used for agricultural fields or pasture for horses and oxen, and the forested hills were used for slash-and-burn agriculture, timber, firewood, and wild food.

By the year 600, about 10–15% percent of Japan's landscape had been converted to monoculture cropping of rice (Fig. 6.7), and the population of the islands probably reached 5 million. This growth in population was due in part to continued immigration from continental Asia. Despite the pressure for more agricultural land to sustain this growth, substantial areas of forest remained intact, even in the more populated parts of southern Japan. As the Yayoi culture ended and successor cultures developed, a ruling class arose and became progressively more powerful, exacting taxes in rice. As their wealth grew so did their desire for building large wooden residences, fortifications, and ceremonial buildings. The ruling classes were initially based in northern Kyushu, but by about 500 their center of power was in Honshu around the areas of modern-day Osaka and Kyoto (the Kinai Basin); by the year 800 they had consolidated their hold over Japan as far north as northern Honshu.

During this period, crops and taxes from rural regions were funneled to the centers of power, but in the rural areas population growth was low. Wild game from the forests continued to provide pelts and horns, which were also sent to markets in the growing cities of the Kinai Basin. Human populations were also depleted by frequent outbreaks of disease, especially smallpox. Northeastern Japan remained sparsely settled, probably because of the difficulty of cultivating rice in its cool climate.

Fig. 6.8   The largest wholly wooden building in the world, Daibutsuden at Tōdai-ji in Nara, Japan, completed in AD 752. Nearby forests have been protected since the construction of this Buddhist temple to provide sources of lumber to replace beams within it.

Rural populations remained at similar levels between about 600 and 1250. In contrast, populations in the centers of power, particularly the capitals of Nara and Heian, went through boom and bust cycles as natural resources were depleted intensively in the Kinai Basin to build the cities and temples (Fig. 6.8). The lowlands of the basin were completely deforested while on the hill slopes the old-growth forest was clear-felled to provide timber for construction. The capacity of these forests to produce the size of timbers needed for constructing the great buildings of Heian was probably exhausted before 900. The secondary forests that developed after the old-growth forest had been logged were heavily relied on for firewood for cooking and industries, including the manufacture of ceramics and metal. As a result, tall trees could never regenerate on the hill slopes. The shrubland that did develop was combustible and burned frequently and the hill slopes began to erode. Attempts to conserve the vegetation and soils of the hill slopes proved futile, and the soils that ran off the slopes filled the river mouths, creating the site of today's city of Osaka. The city of Heian began to decay, the wealth and power of its rulers declined, and after 1250, Japan's ruling classes entered a period of civil war.

Despite this political instability, population growth throughout Japan increased rapidly from about 1200 onward (Fig. 6.6). By 1600, the population of Japan was about 15 million. This was spurred by major advances in agricultural practices, which became far more intensive between 1200 and 1700, so that yields of crops increased from the same land area. Increased agricultural production and more intensive land use was the result of the widespread use of mulch as fertilizer, sustained soil fertility, improved methods of wet-rice cultivation in paddies, and more widespread use of tools and animals in agriculture. More hillsides were cultivated, some hills were brought into pasture, and slash-and-burn agriculture was practiced in steep sites that had been previously untouched. As a result, low altitude forests in central to southern Japan were placed under greater human pressure and turned into a mosaic of forests in various stages of regeneration after disturbance. High altitude forests in Honshu, along the Japanese Alps, remained little affected by human activities.

Centuries of civil war among the powerful ruling families ended in 1600 with the triumph of the Tokugawa faction, which consolidated power throughout Japan. Initially the centers of power were widely scattered in fortified small towns, and these towns were enlarged with the construction of castles, temples, and ships. This construction, as well as the emphasis on agricultural expansion, put further pressure on Japan's remaining forests throughout the country except on Hokkaido. New types of saws and the use of winches and animals to drag logs resulted in forests being logged in remote mountainous areas (Box 6.1 and Fig. 6.9).

Box 6.1. Great forests on a remote island

Yakushima is an island in the northern group of the Ryukyu Islands. It is roughly circular in shape, about 30 km in diameter, and is completely composed of granite. The island rises steeply from the coast to its highest peak at 1934 m where alpine plants grow. The island has steep ravines cut by high rainfall – up to 12 m per year at the summits. On small coastal plains people cultivate crops including sugar cane, sweet potatoes, pumpkins, and tea. Nearly the entire steep interior of the island is forested (Fig. 6.9). In the seventeenth century, the samurai rulers of the south of Kyushu turned their attention to the forests of this island and in particular to the sugi (or Japanese cedar) trees. Some of these trees are huge and ancient – some still standing are more than 5 m in diameter and probably more than 2000 years old. The rulers from Kyushu imposed harsh taxes on the people of Yakushima, who were obliged to fell the sugi forests to meet these taxes. The felling of great sugi trees continued on the island almost until the end of the twentieth

---

Box 6.1. (cont.)

century. In the early 1990s, I carried out research in the warm temperate forests in the Segire River Valley (Fig. 6.9), and camped beside the river. The local people told of their ancestors escaping in remote valleys like this one to hide from the samurai overlords. Certainly my campsite, which was surrounded by tall camelias and related *Stewartia* trees with mottled dark brown trunks, was a beautiful place. The river ran clear nearby. Not a bad place to escape to. The sugi forests in this valley were too inaccessible to have been much affected during the period of logging. However, many of the natural forests of the low slopes had been felled and replaced with plantations of sugi. The extraordinary natural forests of this island have been mostly protected within a UNESCO World Heritage site since 1993. The swap of logging for tourism based on these forests, which also support unique subspecies of Japanese macaque monkeys (Fig. 6.10) and sika deer, has brought its own problems. The narrow road around the remote, forested west side of the island was re-engineered and widened for the expected increase in tourism. PJB

By 1650, supplies of timber were becoming scarce and the deforestation of hillsides caused landslides and flooding. Urban growth stalled and the central government moved to arrest the damage and better manage the forests. From 1700 onward, tree planting and reforestation were instituted and Japan's forests were managed to supply timber, firewood, and wild foods, as well as to protect slopes from erosion. Thus, for the second but not the last time in its history, the Japanese learned the cost of deforestation of its steep hillsides and the need to reforest to protect them. The historian Conrad Totman considered that by 1850, Japan had achieved a highly successful system of long-term sustainable forest management.

The pressure placed on Japanese forests after 1600 and the response to their degradation occurred during a further surge in population growth from about 15 million people in 1600 to about 26 million in 1720. During this period, the last areas of cultivable land for agriculture were exploited and the communal forms of agriculture practiced by villages that had endured for centuries were replaced by individual farmers producing for urban markets. Agricultural activity intensified, based principally on rice but with a wide variety of other vegetable crops, using methods developed

Fig. 6.9   Forests on Yakushima, Japan. The most common tall tree in this rainforest, isunoki (*Distylium racemosum*), is related to witch hazels (see Box 6.1).

over previous centuries (Fig. 6.11). In contrast to European agriculture, there was far less emphasis on raising livestock, other than for cattle to be used for plowing. Human waste was used as an important source of manure in agriculture. People depended on fresh and dried fish for protein much more than in contemporary Europe, but game was still hunted in Japanese forests. Famines during the 1780s and the 1830s, when rice crops failed, caused death and population declines in both the cities and the countryside. The increasingly urbanized population of Japan now became involved in manufacturing as well as commerce. The boom in manufacturing relied on an expansion of mining activities for copper, iron, lead, gold, and silver throughout Japan. Yet within the cities, especially the burgeoning city of Tokyo, care was taken during the Tokugawa shogunate not to further despoil natural resources. Until the late nineteenth century, the rivers and canals of Tokyo were clean and renowned for their delicacies including edible seaweeds. Conrad Totman described how

Fig. 6.10   A subspecies of Japanese macaque monkey that is endemic to Yakushima, Japan (see Box 6.1).

the city's "waste wood and paper products became fuel; the resulting ashes, along with night soil and food byproducts, became fertilizer". This state of affairs did not last into the twentieth century, when the Tokyo Bay became heavily polluted from waste from factories and mining.

Until the 1850s, trade with other nations such as China and Korea, and later with European powers, was on terms dictated by the Japanese leadership. The only attempts at invasions of Japan, by the Mongols in the thirteenth century, failed (see Box 3.2). Trade outposts at cities such as Nagasaki were the points where non-native plant species, such as sugar cane and sweet potato, were brought to Japan. From the 1850s, mounting pressure by foreign powers, especially the USA, finally ended the traditional leadership of Japan and its isolation from the rest of the world. From 1868, the new Meiji government accelerated industrialization as Japan rapidly became a trading nation, and this resulted in greater urbanization and even greater pressure on the islands' natural resources.

Fig. 6.11   Mixed vegetable crops in Yakushima, Japan.

The Meiji government imported new technologies from around the world and began entirely new forms of agriculture. For example, new settlers deforested much of the northern island of Hokkaido in the late 1800s to make way for pasture for dairy farming and for raising sheep, with instruction from agronomists brought from the USA. The hunter-gatherer way of life of Hokkaido's indigenous Ainu people was undermined during the Meiji era under a policy of assimilation and appropriation of land. Another casualty of the expansion of agriculture in Hokkaido was its subspecies of wolf, which became extinct in 1889. The subspecies of wolf that was endemic to Honshu, Shikoku, and Kyushu became extinct shortly afterward in 1905. The loss of a key predator plus reduced hunting by an increasingly urbanized population led to an explosion in populations of the native sika deer (Plate 13). The species composition in many modern Japanese forests has been altered by sika deer browsing.

The Meiji era also saw the expansion of settlements to some of Japan's most remote islands. The Ogasawara (or Bonin) Islands are oceanic islands that lie about 1000 km south of Tokyo. Japanese navigators discovered the islands in the sixteenth century but they remained uninhabited until the nineteenth century. Fishermen and whalers from the USA were the first to colonize the islands but permanent settlement by Japanese people occurred during the Meiji era. The introduction of predatory rats, goats, pigs, and deer destroyed the unique vegetation of the islands. Pigs released on one island multiplied rapidly, preying upon the eggs and young of green turtles that had been previously abundant. Extinctions resulted swiftly. For example, on the island of Chichijima, 13 species of snails unique to the island became extinct since the settlement in the Meiji era, probably as a result of the introduction of ship rats. Laysan albatrosses and short-tailed albatrosses were exterminated from other islands in the archipelago, slaughtered by people who collected their feathers and eggs. On other remote islands in northern Japan, red and Arctic foxes were introduced to develop a fur trade and these animals devastated the colonies of breeding seabirds.

On the main islands of Japan, the rapid rate of urbanization and increase in Japan's manufacturing output during the Meiji era placed huge demands on its forests. Demand for firewood was intense as Japan's population reached about 50 million in 1910. Demand also increased for timber for construction of homes, ships, railways, and as fuel for driving steam power and smelting metal. The Meiji government overthrew the village-based communal ownership of land and forests, and a regime of regionally derived tributes and taxes was replaced by direct governmental control or private ownership of many forests. The consequence of these changes of tenure was an end to traditional forms of land management. By the late 1800s, forests were rapidly felled throughout Japan and widespread erosion and flooding resulted (Fig. 6.12). By the early twentieth century, the government moved swiftly to avert further damage and to ensure that Japan would have timber and firewood for the future. A program of reforestation was initiated, for the third time in Japan's history. However, Japan's disastrous course toward militarization and, ultimately, war in the 1930s and 1940s was to undo these repairs. Japan lacks any substantial energy reserves, such as coal or petrochemicals, and its emphasis on war and conquest was in part motivated by its need for energy and raw materials. War preparations placed heavy demands on timber and fuel and by 1945 all of Japan's most accessible forests, including trees in city parks, had been stripped of timber or scavenged for firewood. After the bombing of its cities and the end of World War II, the

Fig. 6.12   Dam built by Dutch engineers in the Meiji era (late 1800s) to counteract extreme erosion in the granite Tanakami Hills, Shiga, Japan. These hills were deforested to provide construction material for Kyoto.

few remaining forests were placed under pressure to supply wood for urban reconstruction.

6.4 CANARY ISLANDS: FROM SELF-SUFFICIENCY TO TRADING POST AND CASH CROPS

By 1400, the indigenous Guanches had reached all the Canary Islands and established agricultural and fishing communities on them. The Guanches had also either cleared or drastically altered the composition of the islands' forests to make way for grain crops or for herding sheep and goats. Despite the Guanches' North African origins, there is no evidence that they returned, or that visitors from North and sub-Saharan Africa were frequent, or that they affected either Guanche culture or the islands' natural history significantly. An account of a visit to the islands in about 1150 by Muhammad al-Idrisi, an Andalusian Arab, describes a Guanche villager who spoke Arabic, which suggests there must have been some interchange with people in adjacent Africa. During the next two centuries, the islands were visited by Genoans, Portuguese, and Castilians, but again there is little evidence of ecological or cultural impacts of these visitors on the islands or on their Guanche populations.

In 1402, the isolation ended, when Spain began a program of colonization of the islands, swiftly conquering and subjugating the

Guanche populations of the easternmost islands of Lanzarote and Fuerteventura. Spanish conquest of the other islands met armed resistance from the inhabitants. El Hierro was the next island to be conquered, then La Gomera, Gran Canaria, La Palma, and last, Tenerife, conquered after fierce resistance in 1496. Massacre by the Spanish and disease swiftly reduced the Guanche populations, many survivors were forced off their land and into slavery, and the Guanche culture was all but annihilated under a policy of assimilation. A few of the Guanches maintained a lifestyle of nomadic shepherding in the islands' mountains after Spanish conquest.

Settlers from Spain followed soon after the islands had been conquered. The Canary Islands became a vital staging post in Spain's conquest of the Americas, and the settlers of the islands grew rich supplying the Spanish fleets with food, livestock, and goods needed for the voyages. Forests in the interior of Tenerife provided timber for the repair of boats and for the construction of a fishing fleet. Accompanying the Spanish conquerors and settlers were their unwanted cargo of ship rats and house mice. This was a second introduction of house mice; they had already been introduced from Africa by the Guanches. Competition, predation, and disease probably caused the extinction of the lava mouse, unique to Fuerteventura and Lanzarote, soon after Spanish settlement. Norway rats were stowaways aboard ships and arrived sometime in the eighteenth century. Both ship rats and Norway rats (Fig. 6.4) are now on all the main Canary Islands where they are significant predators of the islands' unique pigeons. The rats also feed on several of the large-seeded native trees, and influence the composition of the islands' remaining forests.

Under the Spanish, the landscape of the Canary Islands was transformed to support vastly different agricultural crops from those grown by the Guanches. No longer were the island's crops grown principally to feed the inhabitants but instead the new crops were export commodities destined for Spain and other European countries. The islands in return were dependent on imported goods, including food in the forms of salted cod, ham, and beef, grains such as rice and corn, and textiles. To make way for agriculture and new towns and settlements, and to provide timber for their construction, the settlers cleared or logged nearly all the islands' remaining mid-altitude forests.

Many of the export crops grown on the islands over the next centuries followed boom and bust cycles. Sugar cane, a grass probably originally native to New Guinea but domesticated throughout southern Asia, was the first of the export crops. Sugar cane was introduced to the Canary Islands from the nearby island of Madeira in 1480, was grown extensively, and sustained an economic boom for much of the sixteenth century. Large areas were planted with sugar cane, and fuel was needed to boil the cane juice so that it could be exported, for example, as molasses. On Tenerife, the source of fuel was the high altitude pine forests, which were logged. The slopes laid bare either by sugar cultivation or by forest clearance were then subject to erosion. The cultivation of sugar cane was highly labor-intensive and the Spanish conquerors of the Canary Islands brought in slaves from Africa to work in the cane fields. The sugar boom on the Canary Islands ended in the late sixteenth century when sugar cultivation began on a much greater scale in the Caribbean (see section 6.5).

The next of the "boom and bust" crops of the Canary Islands was grapes, for wine production and export. Grapes were grown mostly on Gran Canaria and Tenerife. The Canary Island wines found particular favor in England, and dominated the English market for the latter part of the seventeenth century. Wine production and export from the islands declined sharply in the late seventeenth century for a combination of reasons: plagues of locusts from adjacent Africa destroyed vines – afflicting agriculture just as it had for the Guanches (see Chapter 5) and competition from other sources, as well as war between England and Spain. The wine industry declined even more during the next century, leading to widespread economic ruin for the residents of the Canary Islands.

From the late 1700s, the third major crop raised was prickly-pear cactuses, native to Mexico. The cactuses were grown as host plants on which to raise cochineal scale insects, native to Central and South America; the insects were used to make carmine dye. The Canary Islands were one of the world's major suppliers of the dye until the 1870s, when the dye was produced synthetically on an industrial basis in Europe; this caused the collapse of the cochineal trade. This collapse resulted in large-scale emigration from the Canary Islands, particularly to Latin America.

Bananas were the fourth export crop to go through a boom and bust cycle. Industrial scale production of bananas began in 1855, with transformation of steep landscapes on La Gomera into terraces for banana production (Fig. 6.13 and Plate 14). Many of the Canary Island bananas were exported to England – the Canary Wharf in London was one of the main points at which they were unloaded. Banana production peaked in 1913, and the trade collapsed with the outbreak of World War I. The collapse of the banana trade prompted a second

Fig. 6.13   (A) Houses and irrigated stone terraces, with cultivation of
bananas, Hermigua, La Gomera, Canary Islands; (B) *Musa*, the common
banana. In Tonga and Jamaica, both green and ripe bananas are eaten.

Fig. 6.14   Pico del Teide, with La Laguna and Las Mercedes in the
foreground, from the southern slopes of the Anaga Mountains,
Tenerife, Canary Islands. The basin where the modern cities are located
was once a forest of oaks and hornbeams, trees that are now extinct
in the Canary Islands. These forests were cleared by the Guanches, the
first people to inhabit this island.

wave of emigration, for example, from La Gomera to Tenerife and
also to Venezuela. Many Canary Islanders emigrated to Puerto Rico
during the late nineteenth century. By the 1920s, many rural villages
in the Canary Islands were deserted and emigration continued as
the islands' economic fortunes declined still further beginning with
the Spanish Civil War in 1936 and subsequent fascist dictatorship in
Spain. Although rural populations declined through emigration, the
populations of the major towns and cities grew and the population
became more urbanized. Thus, despite the collapses of the cochineal
and banana industries, between 1842 and 1940 the population of the
Canary Islands nearly tripled from 240 000 to 690 000 (Fig. 6.14). The
largest city in the Canary Islands, Las Palmas on Gran Canaria, is today
Spain's ninth largest city.

The four boom and bust cycles in agriculture reduced populations of native plants and animals directly through land conversion
and indirectly through deforestation, overgrazing, and the introduction of predators, such as rodents and cats. A once widespread bird,
the Canary Island chat, was confined to Fuerteventura by the early
twentieth century, and a shorebird, the Canarian black oystercatcher,
became extinct by about 1940 after a long period of decline. The fifth

economic boom for the Canary Islands is tourism, which will be discussed in Chapter 7.

6.5 PUERTO RICO AND JAMAICA: CONQUEST, SLAVERY, AND CROPS FOR EMPIRES

By the late 1400s, the Taíno cultures of Puerto Rico and Jamaica had transformed much of the lowlands of both islands into agricultural landscapes and forest fallow. The arrival of the Spanish (1493 to Puerto Rico and the following year to Jamaica), with their diseases and weapons, was a disaster for the indigenous people. Initial encounters between the Taíno and the Spanish were amicable, and the Spanish began using tobacco, the narcotic plant that was native to the islands and cultivated by the Taíno. After the initial encounters, planned settlement of the islands by the Spanish in the early sixteenth century doomed the Taíno. The Taíno resisted the new settlers, but on both islands, the Spanish settlers enslaved them. The indigenous civilizations of both Puerto Rico and Jamaica were all but annihilated by the Spanish conquerors. Some of the crops that the Taíno cultivated, such as cassava, endured and are cultivated up to the present day.

Columbus, on his second voyage to the Caribbean, brought a cargo of plants and animals from Europe, purchased on La Gomera in the Canary Islands. The plants included tree crops such as oranges and limes as well as bananas and plantains; the animals included cattle, sheep, pigs, and chickens – all new to the Caribbean. The Spanish settlers who followed in the next century laid out new townships and fortresses and practiced agriculture on Puerto Rico and Jamaica. Early unintentional introductions to both islands presumably included ship rats. Extinctions of several native animals occurred soon after Spanish arrival. For example, on Puerto Rico, at least one large rodent (13 kg) and an island shrew, which had survived Amerindian settlement, became extinct soon after the Spanish arrived.

The Spanish did not seek to transform the landscapes of either island into large-scale agriculture: both islands served as supply bases and points for repairs and replenishment of ships loaded with gold from Central and South America. Because enslavement of the indigenous Taíno had failed to provide the labor the settlers wanted to produce their agricultural crops, the Spanish next began the transport of slaves from Africa to work the fields. The first slaves arrived in Puerto Rico in 1513. Because Jamaica and Puerto Rico, in contrast to Hispaniola, had little gold, the principal concern of the Spanish was to defend their coasts; total human populations were probably no more than 10 000 on either island a century after conquest.

The lack of strategic value accorded to Jamaica by the Spanish empire was a blunder. The English conquered the island in 1655 and allowed it to be used as a base from which to attack Spanish fleets. The Spanish abandoned the island with little resistance and their African slaves fled to the mountainous interior. There these Africans encountered the surviving Taínos, inter-married with them, and led an autonomous life based on hunting and small-scale cultivation within a largely forested landscape. These residents of the interior put up dogged resistance to British attempts to conquer them over the next 200 years.

The first British settlers of Jamaica grew tobacco, ginger, indigo, and cotton for export, and maize, cassava, and vegetables for themselves. Their main source of meat was green turtles on nearby Grand Cayman. Sea turtles made up most of the meat consumed in Jamaica as late as the 1730s. Jamaica's role from the start of British colonial administration was to be a producer of food for export, mainly to Britain. Likewise, from the outset many of the main foodstuffs that its own population consumed were imported, including wheat flour, salted meat, and fish. Jamaica's dependence on imported staples continues to the present.

The agricultural transformation of the lowlands of Jamaica began in the late seventeenth century with the introduction of sugar cane. The grass grew well in the fertile lowlands of Jamaica, which were completely deforested. The production of sugar was most economical in large estates and the result was that small landowners could not compete and land ownership was quickly concentrated in the hands of a relative few. This scale of production eclipsed that of other exporters, such as sugar cane farmers from the Canary Islands, and ended the Canary Island sugar boom. The export of Jamaican sugar to ports in England and the USA made Jamaica's plantation owners, many of them resident in England, some of the wealthiest of all Britain's colonies throughout the eighteenth century. The remaining lowland forests of Jamaica were quickly felled to make way for sugar cane fields. Tree species that were driven close to extinction included the West Indian mahogany, a tree rapidly over-exploited by furniture makers.

The scale at which sugar cane was grown in Jamaica exacted a terrible human cost. The labor to produce sugar was provided by slaves, mostly from West Africa. Slave capture and transport increased from a small-scale to a wholesale industry, with nearly 4 million people forcibly taken from African ports to British colonies in the Caribbean, the largest of which was Jamaica. Tens of thousands of people died during voyages across the Atlantic. On arrival, the British took slaves for their own colonies but also sold many of them to Spanish colonies, including Puerto Rico. Sugar cane cultivation in Puerto Rico began earlier than in Jamaica but its output was much less during the eighteenth century because sugar cane production from British and French colonies in the Caribbean dominated the trade. As in Jamaica, sugar cultivation in Puerto Rico depended on slavery.

The workforce of slaves in the sugar plantations needed food to sustain them, so various crops were introduced from throughout the tropics to be grown in the small allotments of land that slaves were given. These included African crops, such as yams and the ackee tree. Breadfruit was another tree crop introduced to the Caribbean as food for slaves, brought by the British Admiralty from Tahiti (Box 6.2). The first attempt at transporting breadfruit from Tahiti was aboard the *Bounty*: a mission that failed because of the mutiny aboard while the ship was in Tongan waters off Tofua. Although breadfruit is widely grown and eaten in Jamaica now, the fruit was initially regarded with disdain by the slaves.

Box 6.2. Breadfruit: past and future

Breadfruit has long been a staple of many Pacific cultures. It grows as the fleshy fruit of a tree and has a nutritious and tasty white to yellow, potato-like flesh that can be baked, boiled, steamed, or fried. The breadfruit tree is prolific, potentially producing over 100 kilos of fruit each year, beginning at only three or four years of age. It sounds like the ideal food, so why doesn't everyone have a tree in his own backyard? One reason might be breadfruit's association with slavery. When Captain Bligh set sail for Tahiti in 1789, he was attempting to collect breadfruit to propagate and bring back to the Caribbean to feed African slaves. Bligh was thwarted by the mutiny of his crew on the *HMS Bounty*, but breadfruit did eventually become a staple of the Caribbean slave diet. Another reason breadfruit consumption has declined might be from competition with convenience foods (although what could be more convenient than harvesting dinner from a tree outside your door?). Or perhaps last time you had breadfruit it wasn't ripe enough or cooked correctly. Close relatives of breadfruit (breadnut from Papua New Guinea and

> **Box 6.2.** (cont.)
>
> dugdug from the Marianas Islands) have seeds, but breadfruit is generally seedless. Breadfruit must therefore be grown from cuttings. Early sea-faring Polynesians brought carefully tended cuttings with them on their voyages around the Pacific Ocean. Modern cloning techniques now make the production of young seedlings much more efficient; and efforts are underway to preserve the hundreds of varieties of breadfruit that still survive on Pacific Islands. Next time you get the chance, try some breadfruit chips. You might like them. LRW

Sugar cane fields provided ideal habitat for the ship rats introduced early on by Europeans. Small Asian mongooses were introduced to Jamaica from India in 1872 as a biological control agent against ship rats. These control efforts were unsuccessful; instead the introduction of mongooses proved to be disastrous for the Jamaican rice rat, which became extinct by about 1877. Predation by mongooses probably also caused the extinction of a tree snake and two birds that were unique to the island – the Jamaica petrel and the Jamaican wood rail. Mongooses were also introduced in the late nineteenth century to Puerto Rico but no extinctions of native animals have been directly attributed to the mongoose.

The wealth of the plantation owners of Jamaica was boosted further by the introduction of new crops, in particular coffee, a short tree native to eastern Africa (see Plate 15). Whereas sugar was a crop that grew best in the lowlands, the best quality coffee came from the island's mountainous interior, which was extensively deforested for coffee plantations. The coffee boom lasted from about the 1790s to the 1830s. Once the boom was over, some of the deforested mountainous interior of Jamaica regenerated back to forest cover. Other tree crops were then planted, including Peruvian cinchona that was grown as a source of quinine to combat malaria.

The Jamaican slaves rebelled and were finally emancipated (but not compensated) in 1838. Jamaica's population had grown from an estimated 124 000 in 1700 to 366 000 at emancipation, at which point about 85% of the population were freed slaves. On emancipation, many liberated slaves in Jamaica fled from the sugar plantations where they had been forced to work and moved to the island's forested interior (Plate 5) where they carved out small farming communities. Here they practiced subsistence agriculture, but also cultivated bananas, cocoa, coffee, ginger, and pimento (the dry fruit of a tree native to Jamaica). Their strong reliance on charcoal as a cooking fuel placed pressure on the remaining forests. The people cleared forests to make way for agriculture, cultivating crops in the soils and ash of the felled and burned forests for up to ten years, then clearing a new area of forest and leaving the previous area to regenerate. This process, known as shifting cultivation, typically following slash and burn techniques (see Chapter 5) and continues in many areas to the present.

Sugar production, currently more mechanized, remains important in Jamaica's deforested lowlands, but the prosperity enjoyed by plantation owners diminished after emancipation. Some former sugar plantations in the Jamaican lowlands switched to cattle farms while others began banana production, the first commercial production of bananas in the Americas. Development of the banana industry became possible because of refrigerated shipping, so that Jamaican bananas could be sent in "banana boats" to ports such as Bristol in the UK from the early twentieth century. Jamaica's strong position as a banana exporter waned by the 1940s because of diseases that affected bananas from Jamaica and because of growing competition from other banana-producing countries.

After slavery was abolished in the British colonies of the Caribbean and in the wake of the successful revolution against slavery in Haiti, the Spanish colonies, including Puerto Rico, continued to practice slavery. Former slave owners from other islands immigrated to Puerto Rico. New sugar cane plantations placed increased pressure on the Puerto Rican lowlands and tens of thousands more slaves from Africa were transported to Puerto Rico between 1815 and 1845 to work in the plantations. Thus Puerto Rico went from a small population, 45 000, in 1765 (11% of whom were slaves) to nearly 13 times that number, 580 000, in 1860, of which probably about 40% were slaves. Slavery was finally abolished in Puerto Rico in 1873. The expansion of sugar production in the lowlands displaced the small-scale landowners who had been there. These landowners moved to the unpopulated, mountainous interior of the island. There they created new farms by deforesting the region. The establishment of new populations and towns in the island's interior provided an infrastructure that allowed other crops, particularly coffee, to be developed in the late nineteenth century.

While much of Puerto Rico's agricultural output in the nineteenth century was as crops for export, the amount exported to Spain dwindled rapidly while the food it exported went increasingly to the USA. The USA also supplanted Spain as the main source of imports to the island. In 1898, more than 400 years of Spanish colonial rule

Fig. 6.15   From El Morro, the fortress built by the Spanish from 1539, at the Bahia de San Juan, Puerto Rico, looking along the north-east shoreline.

of Puerto Rico ended when the USA invaded the island during the Spanish-American War. At the end of the war, Spain ceded the island of Puerto Rico and it became a colony of the USA.

Deforestation of Puerto Rico's interior continued after it became a US colony and deforestation reached its peak by about the 1920s. Some of the last remaining forests in the mountainous interior were converted to pasture, croplands, and coffee plantations, and remaining stands of forest were felled for timber and plywood. Reforestation of former pastures began in the 1930s, often with introduced trees such as teak and mango. Since then, mostly native forest species have regenerated around the introduced trees, although sometimes the tree crops still persist.

In both Puerto Rico and Jamaica, the period after slavery during the latter nineteenth and early twentieth centuries also saw the growth of a merchant class, an increase in manufacturing, and the rapid development of cities. In Puerto Rico, the small citadel city of San Juan expanded rapidly beyond its walled fortress on a small peninsula (Fig. 6.15) onto the plains and around its harbor. In Jamaica, the former capital, Spanish Town, became less important as its new capital, Kingston grew quickly. This growth was in spite of several devastating fires and a major earthquake in 1907. San Juan and Kingston became important markets for agricultural produce from the interiors of both islands, and also placed large demands on water resources. As both islands became increasingly urbanized, their populations grew rapidly to 1.9 million in Puerto Rico and 1.2 million in Jamaica by 1940.

6.6 POLYNESIAN ISLANDS, EUROPEAN AGRICULTURE, AND ECOLOGICAL TRANSFORMATION

By the fourteenth century, Polynesians had reached and settled, either temporarily or permanently, almost every volcanic island and atoll within the vast "Polynesian Triangle" of the Pacific, with its corners enveloping the islands of Hawaiʻi, New Zealand (Aotearoa), and Easter Island (Rapanui). The tropical Pacific islands suited to permanent settlement, especially the volcanic islands or the more fertile uplifted atolls, supported dense populations that were maintained by intensive agriculture of starch crops. In the temperate – but much larger – islands of New Zealand, an abundance of bird life sustained initial colonization. The Māori population either cultivated a subset of the tropical starch crops they had brought or learned to use local sources of starch, such as bracken ferns, in the cooler regions where imported tropical crops could not survive.

European navigators reached some of these remote Polynesian islands beginning in the mid-seventeenth century and had reached nearly all of them by the end of the eighteenth century. The societies that Europeans encountered were usually hospitable, as the native protocols demanded considerate treatment of guests. The name bestowed by the English navigator James Cook – the Friendly Islands – to the Haʻapai Group in Tonga recognized the generosity with which the local Tongans provided food and supplies. Polynesian societies had firm regulations about the use of natural resources such as fisheries and forests, including prohibitions that left areas to recover after use. Conflict between the early European navigators and the Polynesians whose islands they landed on was often a result of often unwitting violations of these rules by the visitors. For example, a violation of rules about the felling of trees for new ships' masts in northern New Zealand is a likely reason for conflict that resulted in the death of the French navigator Marion du Fresne. Retaliation from the Europeans, with superior weapons, was swift and resulted in the deaths of many Polynesians.

The European navigators also brought intentional cargo, for example, pigs, dogs, and cats, which they gave to the Polynesians. Although pigs had been introduced by Polynesians to most of the islands that they settled, they either did not survive their voyage of settlement to New Zealand from the eastern Pacific or died soon after arrival. Thus, for the Māori of New Zealand, the arrival of pigs was a welcome source of food. Unfortunately, the pigs also became an important predator on native species of birds and snails. The navigators' less desired cargoes of ship and Norway rats were also deliberately released on many islands – James Cook tied up his ship *Resolution* close to the shore in the southern South Island so that "the vermin could walk ashore over the bridge." The liberation of these rats, along with predators such as cats, was a catastrophe for many native species of animals. In New Zealand, most of its seabird colonies, which had survived the earlier Polynesian introduction of the less voracious Pacific rat, were driven to extinction. James Cook's voyages also introduced potatoes to the Māori and this was an important crop because it could be cultivated in the cooler climates of New Zealand where traditional Māori crops could not.

The European imperial powers, mainly Britain and France, initially considered that most of the remote Polynesian islands were of little value to them, lacking resources of strategic value other than timber for ship building. Moreover, the islands are as far from the European imperial powers as it is possible to get – literally the antipodes – so military conquest of the Polynesian cultures, followed by the maintenance of a far-flung part of their empires for little return, was unappealing. The eighteenth-century voyages, during which Europeans first encountered many of these remote islands, often had an explicit scientific focus. Astronomers and naturalists accompanied most of the early navigators and collected specimens of plants and animals. The specimens that naturalists such as Joseph Banks collected, and the observations that they made, highlighted the unique attributes of plants and animals that occur on remote islands, and their relationships among islands. This, in turn, provided the knowledge that benefited voyages of the early nineteenth century with naturalists aboard – Charles Darwin, Alfred Wallace, Joseph Hooker, and others – who pondered further on the life forms and biogeographic patterns of species that they saw, and used their observations as evidence to support evolutionary theory (see Chapter 4).

The navigators' and naturalists' accounts of some of the natural resources that they saw, especially the abundance of whales and seals, were relayed back to Europe. At the time of first European contact in New Zealand, there may have been 10 000 southern right whales around its coast. Whaling vessels arrived in New Zealand waters in the 1780s, little more than a decade after James Cook's voyages. Among their targets were southern right whales, which were hunted from the 1780s and throughout the next century for their oil and for their baleen, used among other things for making corsets. Whale numbers plummeted as a consequence to less than 500 by 1900. New Zealand fur seals were valued for their pelts in markets in London and China. From about 1800 to 1820, English, Australian, and American sealers killed hundreds of thousands of seals on rookeries around the coast of southern New Zealand and its remote outlying islands to the south. Between 1804 and 1807, about 113 000 skins were taken from the tiny Antipodes Islands alone; the Antipodes lie 820 km southeast of New Zealand. As each new population was discovered, indiscriminate and almost total slaughter followed. The rats and introduced cats introduced by sealers and whalers to the remote islands had disastrous consequences for the remaining animals that were unique to these islands. For example, on the Chatham Islands to the east of New Zealand, Dieffenbach's rail and the Chatham Islands rail became extinct soon after the introduction of these predators. Some of the sealers and whalers settled around the New Zealand coast, and some married into local Māori families. Māori near the coast prospered by supplying the ships with food, including newly introduced crops such as potatoes and wheat.

Fur traders also based themselves in Hawai'i from the late 1780s, using the islands as a staging point to trade sea otter pelts from the west coasts of the USA and Canada to markets in China. The islands were also used as a base for whalers. In Hawai'i, Tonga, and New Zealand, Polynesians traded food for metal implements from the Europeans, including nails and tools, but also guns, which were used to devastating effects in internecine warfare among the Polynesians in each of these island groups.

Following the sealers, whalers, and fur traders were other European settlers, among whom Christian missionaries were prominent. In Tonga, King George Tupou I, with advice from missionaries, made sale of any land to foreigners illegal. This decree was an effective brake against the incipient stages of colonization that were underway in New Zealand and Hawai'i.

6.6.1 The colonial transformation of New Zealand's environment

Merchants established settlements around the coast of New Zealand and some of these became more established townships. Around the settlements, agriculture was begun with imported crops and livestock, such as sheep and cattle. Māori farmed many of the imported crops to supply food to growing towns such as Auckland. The increasing numbers of settlers in New Zealand pressed the British to take a more active role in colonizing the country. The British formally colonized New Zealand by negotiating a treaty with many Māori tribes guaranteeing Māori management of the country's natural resources, including land and fisheries. These guarantees were broken soon afterwards and continued to be broken over the next century.

The formal colonization of New Zealand began a wave of immigration, mostly from the British Isles. Settlements for New Zealand, which were planned in England and Scotland, took no account of the resident Māori. On the South Island, these settlements resulted in the extensive farming of grass crops and grazing of sheep. Uneasy coexistence of Māori with the British immigrants gave way to conflict and finally to war. Between the late 1840s and the early 1860s, British troops were used to take possession of Māori land by force. By the end of the nineteenth century, the Māori had been stripped of nearly all of the land best suited for agriculture. Their populations had been reduced drastically by disease and warfare. Further encroachment by settlers on Māori lands took place in the early twentieth century. Traditional Māori methods of managing natural resources were mostly ignored or negated by the settlers, and were maintained by Māori only in a few remote regions.

The eastern sides of New Zealand's two main islands, which are in a rain shadow, had been transformed into a mostly deforested landscape dominated by long-lived native grasses by fires set soon after Māori settlement (see Chapter 5). European settlers burned these areas with even greater frequency than did the Māori and removed nearly all the remaining forests. The tall, native grasses that could be grazed by sheep and cattle were fired to encourage re-growth of young shoots, but sustained use of fire over decades killed many of these slow-growing grasses. In their place, European pasture grasses were introduced and grew very successfully. The result was wholesale transformation of much of New Zealand into a transplanted European pastoral ecosystem (Box 6.3 and Fig. 6.16). In the wetter parts of New Zealand, Māori had practiced shifting cultivation and preserved much lowland forest for hunting, especially of birds. These lowland forests were felled for timber and many were cut and burned and replaced by European pasture grasses (Fig. 6.17). One example of this was the elimination of the lowland forests of Banks Peninsula in the South Island, which were felled to provide timber for constructing the nearby growing settlement of Christchurch. Once the lowland forests were felled, those on the slopes of the peninsula were burned to make way for pasture grasses. By the beginning of the twentieth century, only 1% of the peninsula's original forests remained. Throughout New Zealand, many of the lowland forests that were felled or burned grew on soils that proved to be entirely unsuitable for agriculture, such as those that supported kauri forests in the northern North Island. Deforestation of the more mountainous interiors continued in the early twentieth century, especially in the 1920s, but many of the farming communities that established there failed within two decades because of the unsuitability of soils for pasture and soil loss due to erosion.

Box 6.3. Driving on the Canterbury Plains

It is a 15-minute drive from the edge of the city of Christchurch to the small town of Lincoln on the Canterbury Plains on the South Island of New Zealand. I have been driving this route to work for much of the past 15 years. The view to the east is of the Banks Peninsula and 100 km west across the Plains, the foothills of the Southern Alps. The snow-capped mountains are plainly visible during clear winter days. In the fields along the drive there is almost nothing to suggest being in New Zealand at all. It is made up almost entirely by the "portmanteau biota" that the historian Alfred Crosby described: the grasses and herbs that made up the fields of England. Those English fields have been reconstituted here, 19 000 km away, and are grazed by sheep and cattle of breeds derived from Europe (Fig. 6.16). The Plains are a windy place and the fields are protected by rows of trees, pruned brutally into rectangles. The trees are all foreigners: Monterey cypresses and radiata pines from California, eucalypts from Australia, and poplars and willows from Europe. Even the stream sides are rank with European herbs. Wild native plants have been all but replaced – one or two specimens of hardy tī or cabbage tree have endured. Flocks of European birds, mostly finches, fly by. The native birds seen on the drive are fairly recent immigrants that flew to New Zealand from Australia – pūkeko (a rail), masked lapwings, welcome swallows, and Australasian harriers. They all thrive in this open farmland. The same imported landscape extends for 100 km to the north, to the south, and to the west, and it is these landscapes that contain the agricultural wealth of the country.

Box 6.3. (cont.)

   Does it matter that many New Zealanders grow up surrounded entirely by a landscape transplanted from the other side of the planet? In the last few years, there have been changes along the route that I drive. Houses have been built in fields where sheep used to graze. Some of the hedgerows, made up of gorse from Europe, that surrounded these fields have been torn out and replaced with mixtures of trees, shrubs, and grasses native to New Zealand. On a little river terrace, children from a local school have planted groves of native trees where there used to be willows. New Zealanders of European descent are beginning to cultivate native plants in the same places from which their forebears so assiduously removed them. New Zealanders are doing this because they have a sense that the country's native plants and animals have declined and will continue to do so unless people make conscious efforts to conserve them. They are also doing this because, generations after their immigrant forebears left the British Isles, they have a sense that a nation founded on a transplanted culture and a portmanteau biota is unsatisfying. Being surrounded by plants and animals unique to New Zealand may result in a less utilitarian attitude to the land than has prevailed and a sense of nationhood in which both the indigenous and the imported are valued. PJB

Fig. 6.16   Cows maintain the Canterbury Plains on Thompson's Track as a grassland composed almost entirely of plants introduced from Europe; in the distance are the Southern Alps of the South Island, New Zealand (see Box 6.3).

Fig. 6.17   Deforestation of southern beech forests in the Catlins, South Island, New Zealand and its replacement by pasture of introduced grass species.

New Zealand's total population grew swiftly. The population increased more than sixfold from 250 000 in 1870 to 1.6 million in 1940. The New Zealand economy was founded on agricultural output that greatly exceeded the needs of its own population: for instance even in 1886 there were more than 16 million sheep raised on New Zealand farms. From the earliest time of European settlement, New Zealand produced food for export, initially to Australia. Once faster, refrigerated shipping began in the 1870s, agricultural commodities, including perishable food such as meat and dairy products, were shipped to Britain and throughout the world. In New Zealand's towns and growing cities, people chose to live in low density housing surrounded by large gardens so that, although populations were small, urban areas tended to be large. By the end of World War I, 60% of New Zealanders lived in towns and cities and that percentage has grown ever since.

New Zealand's agricultural prosperity was achieved through the destruction of the habitats of its native plants and animals. The predatory mammals that settlers introduced had deadly effects on native animals. Ship and Norway rats, and cats were major predators. The last population of a once widespread wren, that may have been flightless, survived on tiny Takapourewa, an islet between the North and South Islands. The wren was driven to extinction in 1895 by cats introduced by lighthouse keepers. Rabbits were introduced to New Zealand as a source of game. Their numbers burgeoned and they competed with livestock for pasture grasses; so, in an attempt at biological control, stoats, weasels, and ferrets were introduced. This attempt was as misguided and ecologically disastrous as the introduction of mongooses had been in Jamaica. Stoats are efficient predators and are one of the main reasons for the rapid extinction of some of New Zealand's unique birds, and the reduction of other species to a few remnant populations on remote islands. Laughing owl and huia (a New Zealand wattlebird) succumbed to predators by about the end of the nineteenth century. Other birds, such as piopio and bush wren, became extinct in the twentieth century; still others, like mohua, are now in danger of extinction.

The European settlers of New Zealand not only transplanted their pastoral landscape but also introduced other European biota with the express intent of making New Zealand more like England. Many European birds were introduced and, among these, blackbirds and chaffinches are some of the commonest birds in New Zealand today. Brown and rainbow trout were introduced for anglers and these predatory fish drove at least one native fish, the grayling, to extinction before 1940. Red deer were introduced as early as 1851 for hunting, and subsequently another six species of deer were introduced (see Chapter 7). These animals have grazed some native plant species in forests to the point that their regeneration is very infrequent. Brushtail possums were introduced from Australia to begin a fur trade and have become widespread, browsing some plants like the native mistletoes until they have become very rare. Brushtail possums also prey upon birds and their eggs. All of these introduced animals eventually have spread to the remotest parts of the North and South Islands. The adverse consequences of some of these introductions were quickly apparent, and this prompted some of the first efforts internationally to eradicate introduced animals. For example, pigs were eradicated from a small island (Aorangi) in 1936 where they were preying upon nesting seabirds, and goats were eradicated from another island (Great Island) in 1946, where they had reduced the populations of two woody plants unique to the island to single individuals (see Box 4.3).

The settlers to New Zealand introduced thousands of new plant species, from garden plants to tree crops, and many of these have escaped and transformed New Zealand's landscapes. The number of introduced plant species that have naturalized (i.e., have self-sustaining wild populations) now exceeds the approximately 2360 native plant species. Some habitats, such as riverbeds, are dominated by introduced plant species. New Zealand's agriculture is almost completely based on introduced plant species, as is its timber industry. The natural forests were harvested for timber until the 1980s: there was clear evidence from early in the twentieth century that growth rates of its native timber trees would never match the rate at which they were felled but this evidence was generally ignored. After the exhaustion of timber from its natural forests, New Zealand became reliant upon faster-growing North American trees such as radiata pine and Douglas fir, and large-scale plantations of these trees were started in the 1930s; New Zealand has no native trees in the pine family. Both radiata pine (see Section 7.3.2) and Douglas fir, along with other introduced species of pine, have since naturalized widely.

6.6.2 Hawai'i joins the global markets

Unlike the Polynesian society of New Zealand, where authority rested with individual tribes rather than with a central leadership, Hawai'i and Tonga were kingdoms and strongly hierarchical. The strong central authority of royalty in Tonga was the means by which colonialism, to a great extent, was resisted, as we discuss later in this chapter. This was not the case in Hawai'i, where the authority of the Hawaiian royalty dwindled and ended in less than a hundred years. The use of Hawai'i as a base for fur traders exporting North American otter fur to China brought attention to a Hawaiian resource desired in China. The native Hawaiian sandalwood trees were desirable in China for the manufacture of furniture and incense. Sandalwood was extensively harvested in low- to mid-altitude forests for export for about 15 years from 1810 (Fig. 6.18). The profits of the sandalwood trade went partly to Hawaiian nobility, who directed labor away from the agricultural base of their society to extract timber for export. The settlement of Hawai'i's coast by whalers also diverted labor away from its traditional base, as Hawaiians instead worked to provide items needed on the whaling ships, such as rope made from coconut fiber, or salted pork for the voyages. As the settlements became larger and more permanent, Hawaiians increasingly moved to them.

At the same time that these events led to the undermining and neglect of traditional Hawaiian agriculture, the settlers used some of the traditional agricultural land for new imported crops and the grazing of imported livestock. The growing community of settlers from the USA and the British Isles began to influence the authority of the Hawaiian kings. In 1848, King Kamehameha III instituted land reforms

Fig. 6.18   Sandalwood was exported from Hawai'i to China and
volumes were calculated to match the hold of ships used for transport.
This hole on Moloka'i represents such a volume.

that granted land to most native Hawaiians but, in contrast to Tonga,
allowed them to sell land to foreigners. The sale of land to foreigners
was widespread; attempts to stop the process came too late.

Once the sale of lands was underway from the late 1840s, foreigners turned Hawai'i into a source of exported cash crops. Sugar
cane, which had already been a boom and bust crop in the Canary
Islands, Jamaica, and Puerto Rico, rose to be the dominant exported
crop from Hawai'i by the 1870s. Its cultivation continued into the
twentieth century and became increasingly reliant on immigrant laborers from China and Japan. The settlers irrigated fields for sugar cane
cultivation in areas where Polynesian agriculture had never been practical. Dry plains, such as the 'Ewa Plain near Honolulu, were made
arable through the digging of wells and importing of top soil from the
uplands. Cultivation of sugar cane on islands such as Lāna'i resulted
in erosion, and the diversion of water from its natural courses killed
stream life. Pineapple (Fig. 6.19) was the next boom and bust crop to
follow sugar; its boom lasting until the late twentieth century (see
Chapter 7), and other crops including coffee followed. The usurping
of traditional royal authority in Hawai'i was complete in 1893 when
the USA abolished the monarchy, and then it formally annexed the
islands in 1898. Sugar cane farmers and other businessmen, seeking

Fig. 6.19   A pineapple plantation on Oʻahu, Hawaiʻi. Pineapple production in Hawaiʻi has declined sharply in recent years due to overseas competition and increased production costs.

more authority over agricultural production, had a central role in the demise of the Hawaiian monarchy.

A cargo of unintended introductions coincided with the European colonization of Hawaiʻi. Although ship rats often invaded Pacific Islands soon after the first Europeans arrived, in Hawaiʻi their arrival may have been as late as the 1870s. When they did arrive their effects were as devastating as they had been on many other islands. Ship rats have now reached most of the islands of Hawaiʻi and they are one of the most likely causes of extinctions among the birds unique to Hawaiʻi, including koa finches found only in the lowlands of Kauaʻi, Oʻahu, and Maui. Two species of koa finch occurred in the koa forests in the uplands of the Island of Hawaiʻi. Both became extinct in the first half of the twentieth century. It seems most likely that both species were the last vestiges of populations whose main habitat in the lowlands had already been destroyed by Polynesians (see Chapter 5). The tipping point for these birds was destruction and degradation of their last forest habitat in the uplands after European settlement, coupled with predation by the ship rats. Ship rats continue to be a major problem for conserving Hawaiʻi's remaining native birds; they are one main contributor to the on-going decline of species such as ʻelepaio on Oʻahu and palila on the Island of Hawaiʻi. Ship rats were also a major economic problem, especially in the sugar cane fields. Their outbreaks prompted the same futile attempt at biological control – using mongooses – as was attempted in Jamaica. It was from Jamaica that mongooses were imported to the Island of Hawai'i in 1883 and then shipped to plantations on the other large islands, where they are significant predators of native birds, as they are in Jamaica. Kaua'i and Lāna'i remain free of mongooses. Although mongooses were shipped to Kaua'i they never reached land. An account of the fate of the mongooses destined for Kaua'i is that one of the mongooses bit a dockworker who, enraged, flung all of the caged animals into the harbor to drown.

Europeans did not introduce just predatory mammals to Hawai'i. They also released several species of browsing and grazing animals, of which goats were the most widely liberated across the islands of Hawai'i, and these became abundant in many habitats from lowlands to high mountains, and from wet habitats to dry. The effects of introduced herbivorous animals on forests have been as destructive as on other islands worldwide. Browsing and grazing animals selectively eliminate certain species, prevent regeneration of native plants, and maintain large areas in grassland. Their destructive effects were recognized early on, and herbivorous animals were eradicated from Ni'ihau in 1910, and from other larger islands (Lāna'i and Kaho'olawe) in the 1980s. Other browsing and grazing animals introduced to various islands in Hawai'i during the nineteenth century included mouflon and rabbits from Europe, and axis deer from India (see Section 7.5.2).

Since European contact, bird diseases transmitted by mosquitoes, which were inadvertently introduced in 1826, have also had dire effects on Hawai'i's unique birds. Avian pox began to affect birds such as the 'ō'ō from the 1880s. Different species of 'ō'ō became extinct on O'ahu, Maui, Moloka'i, and then Lāna'i between 1899 and 1931. Avian malaria began to affect Hawai'i's native birds from about 1940 and has been implicated as the cause of likely extinction of the last species of 'ō'ō, on Kaua'i, in about 1987 (see Chapter 4).

Thousands of plant species have been introduced to the islands of Hawai'i over the 200 years since European contact and settlement. Some of these plants have grown so successfully in Hawai'i that they now dominate many of the landscapes of the islands (Fig. 6.20), especially in the lowlands and the dry, leeward sides of the islands. For example, large parts of the Island of Hawai'i have been transformed into pasture entirely dominated by introduced grasses. Some of the introduced plants are functionally different from any native species

Fig. 6.20   Invasive strawberry guava in Hawai'i. Strawberry guava invades native forests in Hawai'i where it forms very dense thickets. It often out competes native trees because it can sprout from its root and its seeds are spread rapidly by many types of animals.

and can be agents that cause further change. For example, several grasses introduced to Hawai'i from East Africa as forage for cattle are highly combustible, but very few of Hawai'i's native plants tolerate fire. When the introduced grasses burn they promote their own re-growth and this situation results in their dominance because fires successively eliminate any remaining native trees that may have survived in these grasslands. When pastures of these introduced grasses adjoin native forests, fires through the grasses scorch the forest margins, the trees die, the grasses invade, and successive fires push the forest boundaries back further. Other introduced plants invade disturbed sites, such as bare lava formed by the active volcanoes on the Island of Hawai'i. One of these is fire tree, a tree native to the Canary Islands (Fig. 6.21 and Box 6.4).

Box 6.4.   An invasive plant alters ecosystems

Fire tree was introduced to the Island of Hawai'i in the nineteenth century by Portuguese settlers from the Azores and Madeira Islands in the Atlantic Ocean. The Portuguese settlers used this familiar tree for erosion control in lowland sugar plantations and in some places it developed dense forests. In 1961, one individual was planted in the town of Volcano at 1000 m

---

Box 6.4. (cont.)

elevation near areas with recent volcanic activity. Native plants
in this area grow in very young, nitrogen-poor soils. Fire tree
has symbiotic, nitrogen-fixing bacteria in root nodules that give
it a competitive advantage because it is relatively well-supplied
with nitrogen while the native plants are not. Fire tree dispersal
is aided by a non-native bird, the Japanese white-eye. Fire tree
spread rapidly and by 1985 there were tens of thousands of
individuals in the region, despite over a decade of removal
efforts (Fig. 6.21). I worked to understand how a population of
trees could spread so widely and what its effects were on native
plants and soils. A parasitic insect was introduced from fire
tree's native habitat (a two-spotted leaf roller) in the 1990s, but
it only made modest reductions in fire tree populations and even
developed its own parasite over time. Currently, fire tree is still
widespread, is still being controlled in certain areas, and appears
to be more sensitive to acid rain from renewed volcanic activity
than the native trees. However, because the leaf litter of fire tree
decomposes to form soils that are rich in nitrogen, this in turn
favors colonization by other non-native species of plants. LRW

Since European contact, an extraordinary 142 bird species have
been introduced to Hawai'i from six continents and 54 of these have
formed breeding populations. While many of the native bird species
have become extinct or very rare, and others are more or less confined
to the much-reduced area of natural forests, the non-native (introduced)
birds dominate much of the lowlands. These are the zones where non-
native plants are often abundant and introduced birds can play important roles in their ecology. The ecologist Dan Simberloff coined the
phrase "invasional meltdown" to describe a process where interactions
between introduced species favor each other's growing dominance.
There are many examples of this process in action in Hawai'i. One is
from a recent study in dry forests on the leeward side of the island
of Maui where the bird community is comprised solely of non-native
species. The Japanese white-eye is the most common of these birds,
and it disperses seeds of mostly non-native shrubs into these forests. In
contrast, fruits of most of the native plants in the forests are scarcely
dispersed away from parents. The prognosis for these native forests,
without human control of invasive non-native plants, is grim.

Fig. 6.21   (A) Large, mature fire tree (*Myrica faya*) with multiple trunks formed by sprouts, native to warm temperate rainforest near Contadero, Garajonay National Park, La Gomera, Canary Islands; (B) The invasive, nitrogen-fixing fire tree on the rim of Halema'uma'u Crater, Kīlauea Volcano, Island of Hawai'i.

6.6.3 Tonga retains local control of land

Compared with Hawai'i and New Zealand, Tonga's society, its landscapes, and its animals and plants were far less transformed after
European contact during the period up to 1950. Direct colonization
of Tonga by Europeans was countered effectively by King George
Tupou I during the nineteenth century, so that today Tonga's population is overwhelmingly Polynesian in its makeup. However, Tonga's
Polynesian society changed to a large degree in response to contact
with Europeans. For example, there was near-universal adoption of
Christianity early in the nineteenth century, and serfdom was abolished in 1862. Subsequent land reforms allotted all Tongan males
over 16 years of age parcels of land of 3.34 hectares on which to grow
crops: a system still in place now, even though the bank of unallo-
cated land is dwindling rapidly. It became legal to lease, but not sell,
land to non-Tongans, so that a small settlement of Europeans became
established. Tonga entered into treaties with various European powers, including Germany, during the nineteenth century. Its treaty with
Britain in 1900 made it a state under British protection, although it
remained self-governing.

Subsistence agriculture in Tonga has remained the principal
form of land use, as it was before European contact (Fig. 6.22). However,

Fig. 6.22   Local market in Neiafu, the capital of Vava'u, Tonga. Items
displayed include giant taro, yams, bananas, pandanus leaves, and taro
leaves.

Fig. 6.23   Vanilla beans, a valuable export crop from Tonga.

Tongans began to cultivate crops for export. German settlers and traders introduced new crops. For example vanilla, an orchid native to Mexico, has been cultivated as an export crop in Tonga since the 1870s (Fig. 6.23). The Tongan nobility worked with German settlers to lay out coconut plantations throughout the country, so that coconut palms, which grew wild by the shores and had been in home gardens, became a widespread feature of the landscape. The coconuts were grown for the export commodities: fiber, oil, and desiccated coconut.

As in New Zealand and Hawai'i, introduced mammals changed the ecology of Tonga. Introduced goats prevented forest regeneration, and predatory rats and cats resulted in many native plant and animal species becoming rare. For example, the once-widespread shy ground-dove is now more or less confined to islets that lack ship rats. Land birds such as European starlings and red-vented bulbuls were introduced to Tonga and are now common.

6.7 ICELAND FINDS A PATH FORWARD AFTER ENVIRONMENTAL DEGRADATION

By 1800, a little less than 1000 years after the Norse settlement, Iceland's population was only about 50000. Iceland was then an impoverished Danish colony, and its agricultural economy had been undermined because of environmental degradation brought about by overgrazing by sheep. The population of Iceland had suffered through harsh colonial rule. For the previous 200 years, trade in agricultural produce and fish within Iceland, exported goods, and imported goods were all strictly controlled by the Danish colonial administrators. In 1800, the Danes abolished the Icelandic parliament, the Alþingi, the last vestige of Icelandic independence. Yet Danish authority throughout its colonies waned in the following decades. Denmark's trade monopoly with Iceland ended in 1854. Iceland's route to independence began with home rule in 1901. When Denmark was occupied by Germany in 1940, Britain occupied Iceland along with troops from the USA. In 1944, near the end of World War II, Iceland became an independent state.

Iceland's agricultural economy since settlement has depended on raising sheep (see Fig. 5.8). In 1784, sheep numbers were probably at their lowest in centuries (about 50 000) but, by about 1850, this had risen to more than 600 000 and there were still about 400 000 in 1950. This increase in sheep numbers was driven by export trade not only of wool but also of lamb to Europe. The large numbers of sheep that the country supported during its path to modern independence exacerbated the problems of soil degradation, which had plagued Iceland since soon after human settlement. The Icelandic home rule government established a Sand Reclamation Institute in 1907 (now its Soil Conservation Service) with a mandate to restore large areas of denuded and mobile soils, especially in its upland regions, and to combat further erosion through promotion of more sustainable methods of agriculture (Fig. 6.24).

Fig. 6.24   Land use in Iceland around Lake Myvatn including agriculture and lava flows. The lake is a bird sanctuary.

Increased wealth from sheep and fishing resulted in a slow increase in Iceland's human population. By 1870, there were about 70 000 Icelanders. During the 1880s many Icelanders emigrated to North America and Brazil, in part due to famine following severe winters and damage to pasture caused by the eruption of the volcano Askja. Icelanders became more prosperous in the early twentieth century as the basis of the economy shifted from agriculture to fisheries, especially after Iceland acquired its first trawlers in 1905, a fishing fleet that grew to 40 by 1930. The proportion of the population engaged in agriculture dropped from about 80% in 1890 to about 40% in 1920. After 1890, fewer Icelanders emigrated but instead left small inland and coastal fishing hamlets to move to the capital city of Reykjavík; Iceland therefore became increasingly urbanized. Reykjavík's population increased rapidly from 1150 in 1850 to 11 500 in 1910 to 56 000 in 1950; by then one-third of Iceland's total population of 144 000 lived in Reykjavík.

6.8 COMMON TRENDS

The nine island groups we consider in this book span gradients of isolation and age (see Chapter 2). The time that humans have been present on the islands ranges from hundreds of thousands of years in the British Isles to mere hundreds of years in Hawai'i, Iceland, and New Zealand (see Chapter 5). Nonetheless, there are remarkably consistent themes across the islands in terms of how humans have modified the landscape. In the British Isles and Japan, with long-established human populations, the early hunter-gatherer civilizations had low populations and did not disrupt the forest cover on the islands. They did manipulate the forest composition by cutting it for fuel and burning it to provide habitat suitable for wild plants and animals to hunt. On both islands, the development of agriculture led to the replacement of forest cover to make way for crops. The introduction of cereal crops from the nearby continent (wheat and other grains to the British Isles and rice to Japan) required domestic animals to cultivate the land. Crops resulted in a transition to a more settled life style and enabled the population to grow. Once metal technology was developed in these islands, wholesale cutting of lowland forest and plowing of soils ensued in both countries, and on the British Isles, where much of the landscape is relatively flat and accessible for agriculture, the result was almost complete deforestation within 2000 years.

In contrast to the British Isles and Japan, the first humans to the other seven island groups we consider all arrived with crops and the appropriate agricultural skills to cultivate them. Those settlers cleared extensive areas to cultivate crops but many forested areas remained (see Chapter 5). When new settlers arrived from Europe to all of these islands, their treatment of the original cultures was often brutal and the original crops and management of the environment mostly abandoned. Exceptions include Iceland, where the original crops were maintained, and Tonga, where reliance on Polynesian crops continued even after European contact. The new technology and agricultural practices brought by Europeans were often less oriented toward feeding the islands' inhabitants than providing crops for distant markets. The result was often wholesale deforestation, and unsustainable production of cash crops that went through boom and bust cycles. Sugar cane was one of these crops that relied on cheap land and labor and was a reason for deforestation of the lowlands of the Canary Islands in the sixteenth century, Jamaica in the next two centuries, and Puerto Rico and Hawai'i in the nineteenth century, as each supplanted the former with cheaper labor.

In the British Isles and Japan, population growth was sustained by increasing conversion of land to agriculture or by increasing mechanization. Urban populations began to grow in cities, placing demands on natural resources such as timber for construction and fuel. The Japanese learned the need for sustainable management of steep hillsides, because, once stripped of trees, widespread erosion took place. In contrast, less concern was paid to land management on islands such as Jamaica where absentee English landlords often focused more on profit than on learning the intricacies of sustainable agriculture where climates and soils were unfamiliar to them. Unsustainable agriculture and environmental degradation on some of these islands, such as Iceland and the Canary Islands, resulted in human suffering and emigration. Emigration, in turn, sometimes led to reforestation, as in Puerto Rico.

The loss of forests resulted in extinctions of native plants and animals throughout the islands. These included large mammals in the British Isles and Japan through loss of habitat and direct hunting. Rats were particularly devastating predators on native plants and animals and caused extinctions of many of the more unusual plants and animals that had survived the first settlement of the other seven island groups. The net result of the transport of plants and animals worldwide was to erase many of the most distinctive native species and to replace them with an increasingly homogeneous group of species worldwide, not only pasture and crop plants, but also escaped garden plants, birds, and mammals. The lowland plants and animals of islands as far apart as Jamaica and Hawai'i are now remarkably similar. They are comprised of plants that can survive frequent fire and grazing, and animals that can cope with predatory mammals. Native plants that were poorly defended against herbivores, and large flightless birds, have become extinct on many islands worldwide. Translocations have occurred between islands and some of the plants and animals from one island have become significant components of another islands' ecosystems. For example, birds introduced from the British Isles are among the commonest birds in New Zealand, and fire tree (from the Azores) and Japanese white-eye (from Japan) are now important components of lowland forests on the Island of Hawai'i.

By 1950, the trend to increasing urbanization that began in the British Isles and Japan was underway in all the island groups we consider, even in still largely agricultural Tonga. Urbanization placed new pressures on island environments to support industries, such as the provision of energy and raw materials. Rapidly growing populations required increased food production, at the expense of island environments. Some of these food demands were met with imported produce. Imperial wars were fought to attempt to secure resources for islands. Clearing of land for new cash crops on some islands penetrated further into the little remaining land that was still covered with natural forests. All of these trends continued at an ever-increasing rate in the latter part of the twentieth century, as we will see in Chapter 7.

SELECTED READING

Allen R.B. and W.G. Lee (eds.) (2006). *Biological Invasions in New Zealand*. Berlin: Springer.
Campbell, I.C. (2001). *Island Kingdom: Tonga Ancient and Modern*, 2nd edn. Christchurch: Canterbury University Press.
Crosby, A.W. (2004). *Ecological Imperialism: The Biological Expansion of Europe, 900–1900*, 2nd edn. Cambridge: Cambridge University Press.
Encyclopedia of Puerto Rico (n.d.) http://www.enciclopediapr.org/ing/ (accessed 13 April 2010).
Fernández-Palacios, J.M. and R.J. Whittaker (2008). The Canaries: an important biogeographical meeting place. *Journal of Biogeography* **35**: 379–387.
Grove, R.H. (1995). *Green Imperialism: Colonial Expansion, Tropical Island Edens and the Origins of Environmentalism, 1600–1860*. Cambridge: Cambridge University Press.
Kristinsson, V. (1973). Population distribution and standard of living in Iceland. *Geoforum* **4**: 53–62.

Martin, J. (1817, reprinted 1991). *Tonga Islands: William Mariner's Account*. Nuku'alofa: Vava'u Press.

Park, G. (1995). *Ngā Uruora: The Groves of Life*. Wellington: Victoria University Press.

Rackham, O. (1986). *The History of the Countryside*. London: Dent.

Richards, J.F. (2005). *The Unending Frontier: An Environmental History of the Early Modern World*. Berkeley, CA: University of California Press.

Roberts, N. (1998). *The Holocene: an Environmental History*, 2nd edn. Oxford: Blackwell.

Totman, C.D. (2000). *A History of Japan*. Oxford: Blackwell.

Ziegler, A.C. (2002). *Hawaiian Natural History, Ecology, and Evolution*. Honolulu, HI: University of Hawai'i Press.

7

Islands in the modern world, 1950–2000

7.1 INTRODUCTION

In this chapter, we explore the environmental consequences of human habitation on the nine island groups during the period from approximately 1950 to 2000. We also examine the reciprocal impacts of those islands' environments on human activities. The loss of life during World War II served to reset many national priorities and the second half of the twentieth century became a time of great technological development, wealth creation, and population growth. On the whole, this half-century was one of optimism. Advances in technology permitted more intensive and extensive exploitation of fisheries, agricultural land, mines, and military reservations. Each of these advances had its own ecological consequences. Wealth brought leisure and rapid expansion of recreational hunting and tourism to islands, again with significant impacts on the native flora and fauna as well as on natural resources such as clean water and air. Ample energy and resources, improved transportation, plus optimism about the future led to fast growth of human populations and the concentration of people in urban areas on many islands. However, for some more remote islands, the advent of rapid transport led to population decline as people moved to urban areas elsewhere. These advances in technology and increased human control over the environment led to a rapid transition from native-dominated ecosystems to those that are a mix of native and introduced species and to some that are largely created by humans. Urban ecology is now a field of research that addresses which types of species withstand the intense human impacts of urban ecosystems and which kinds often become extinct. We explore how human-caused species introductions on the islands have altered the structure and dynamics of biological systems. Finally, in this chapter, we examine how humans have responded to such rapid technological, demographic, and economic changes, both by conserving and restoring ecosystems and how human activities are shaped by island environments. In the final chapter, we will explore the current environmental status of the islands and how ecological limits are becoming widely recognized.

7.2 STATE OF THE ENVIRONMENT IN 1950

In 1950, there was a cautious optimism in the world as the specter of World War II began to fade. Island groups most directly impacted by the war (Japan and the British Isles were both extensively bombed) restructured their economies, started to heal wounds, and encouraged population growth. However, it was optimism tempered by the presence of other international conflicts such as the nuclear arms race and wars in Korea and Southeast Asia. The direct environmental damage from World War II was profound but complex. For example, in Japan, populations of fur seals and song birds declined and old growth forests were logged, but fisheries recovered due to a lack of fishing boats. Japan also suffered much environmental damage from fire storms that resulted from bombing and radiation from nuclear bombs exploded over Hiroshima and Nagasaki. Indirect consequences for the environment due to war activities on the most affected island groups included increased flooding and erosion following deforestation from bombing. Despite the tremendous economic and environmental damage, commercial markets were re-established within and among nations, often at larger scales than before World War II. Natural resources continued to be considered as commodities to extract and sell as they had been for centuries on some islands (see Chapter 6). Agriculture was just beginning its great leap forward with improvements in technology and increased availability of chemical fertilizers. Improved agricultural conditions led to the establishment or expansion of crops. Improved air and boat travel greatly facilitated the export of natural resources from islands and the import of machinery and tools to extract those resources. Expansion of manufacturing industries increased the demand on natural resources. Some island groups, such as the British Isles and Japan, began to focus on information technology and the development of vast financial sectors. Globalization coupled with cheap labor and low environmental standards shifted manufacturing plants to developing countries. However, increased island population and continued resource extraction (such

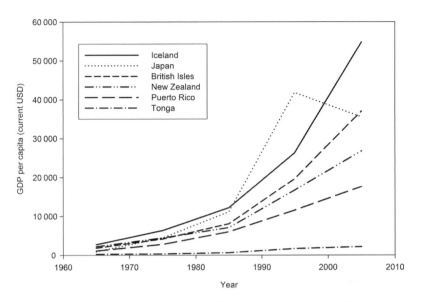

Fig. 7.1   Gross domestic product (GDP) per capita in current US dollars (USD).

as coal) meant that even in the British Isles and Japan, the environmental pressures did not diminish. Islands such as Jamaica continued to be sources for raw materials for manufacturing. The gross domestic product (GDP) of each island group grew, sometimes quite rapidly (Fig. 7.1).

In 1950, access to the most remote of the nine island groups was still limited. Commercial flights were relatively slow, expensive, and infrequent. Island tourism in 1950 was largely limited to wealthy mainlanders who could afford both the expense and the time to reach remote islands. However, domestic travel had begun, with camping and simple cabins or inns providing lodging for travelers by train, bus, or other public transportation in places such as New Zealand and the British Isles.

Urban centers in 1950 included between 13 and 41% of the total population of island groups such as Tonga, Jamaica, Japan, and Puerto Rico while urban centers included 70 to 80% of the population on other island groups (notably New Zealand, Iceland, and the British Isles; Fig. 7.2). In the less urbanized set of island groups, substantial numbers of people still farmed the land and lived in rural villages with limited urban contact. Some cultural groups were still predominantly rural. About 75% of Māori in New Zealand still lived in rural

Fig. 7.2   Main Street, Chester, England.

areas, although some migration to urban centers had begun. Human populations most impacted by war included Japan (that lost 3.78% of its 1939 population), the British Isles (0.94%), New Zealand (0.73%), and Iceland (0.17%). However, population growth rebounded. Improved nutrition, political stability, steady jobs, and other improvements in people's lives led to the birth of the largest number of people in one generation ever recorded (the Baby Boom). This age cohort would go on to dominate cultural, political, and demographic indicators for the rest of the century.

In 1950, the consolidation of popular environmental movements, which focused on global environmental problems, was still at least a decade off. The kernel of a new respect for the land and its products had been suggested by Aldo Leopold in his book *A Sand County Almanac* (published in 1949) and environmental efforts had been successful on many islands prior to the 1950s (such as national park establishment, see Chapter 6 and Table 7.1). The dangers of

Table 7.1. *First national park establishment on the island groups. Puerto Rico does not have a national park. Five additional national parks were established in Japan on 4 December 1934. Those listed were established on 16 March 1934*

Date of establishment	Island group	National park	Principal features
1887	New Zealand	Tongariro	Volcanoes
1916	Hawai'i	Hawai'i Volcanoes	Volcanoes
1930	Iceland	Thingvellir	Historical place
1934	Japan	Kirishima-Yaku, Unzen-Amakusa, Setonaikai	Volcanoes, lakes and islands
1951	British Isles	Peak District	Moorland, bogs
1954	Canary Islands	Teide and Caldera de Taburiente	Volcanoes
1992	Tonga	'Eua	Coastal cliffs, forest
1993	Jamaica	Blue Mountains and John Crow Mountains	Mountains, forest

pesticides (Rachel Carson, *Silent Spring*, published in 1962) and over-population (Paul Ehrlich, *The Population Bomb*, published in 1968) had not yet been presented.

In this chapter, we examine the rapid ecological changes that occurred on the nine island groups during the latter half of the twentieth century. Each of these changes stems from the social, economic, and cultural contexts of the time, which were also undergoing dramatic shifts. Humans are the dominant force in all of these changes. However, humans ultimately rely on natural resources and the services (such as clean air, clean water, and fertile soil) that nature provides. Therefore, human activities are bounded by the condition of those services. Islands provide a clear reminder of how natural resources are limiting because of their obvious geographical and resource finiteness. This half century (1950–2000) thus became a lesson in how humans go from seeing the natural world as a set of resources to be extracted to a growing awareness that ecological limits are real and ultimately control what we can do on both islands and mainlands.

7.3 TECHNOLOGY

7.3.1 Fishing

Despite some overexploitation of fisheries upon initial human contact (see Chapter 5), sustainable levels of fishing have supported islanders for centuries. Many Polynesian cultures built elaborate fish ponds along the coasts as an early form of aquaculture (see Fig. 5.6). Some of these are still in use today, such as on Molokaʻi, Hawaiʻi. However, overexploitation of marine and freshwater resources also occurred whenever human population pressures increased, perhaps prompting some of the Polynesian migrations. Island fisheries often became severely depleted during the colonial period of the last several hundred years, in places where many nations systematically exploited resources, such as crayfish, orange roughy, and snapper in New Zealand coastal waters or cod in the North Atlantic. Improved technology has increased the damage to aquatic ecosystems. Local measures used by fishermen to increase their catch include dynamite and rotenone that kill all the fish on a portion of a reef. Offshore fishing has become more efficient through the use of technological advancements such as steam- and diesel-powered vessels, beam trawls to rake the sea floor, and sonar fish finders. With growing populations to feed and a misplaced optimism that fish stocks were inexhaustible, the era of over-fishing began in the mid-twentieth century.

Globally, fish catches peaked in the 1970s resulting in fish population crashes. Overexploitation based on remote market demands, profit motive, and tenuous feedback loops between supply and demand often resulted in irreversible damage to the resource. Today, many fish populations are gone from large areas of the oceans. Other aquatic organisms have also been affected. For example, populations of Caribbean branching corals have declined an estimated 70% since humans arrived, Caribbean green turtles three- to 16-fold in the last 300 years, and New Zealand oysters ninefold in the last 30 years. People have substantially altered aquatic food webs, and populations of some organisms (whales, sharks, sea cows, fish, turtles, oysters, and coral) have been reduced. There has been a subsequent rearrangement of food chains that has promoted other organisms, including sea urchins, sponges, jelly fish, worms, and phytoplankton. Overharvesting of marine life unfortunately often accompanies further damage from chemical pollution and mechanical destruction of habitat. This damage makes surviving marine organisms more susceptible to diseases and habitats more susceptible to invasion by nonnative species. Aquatic ecosystems have been fundamentally altered by humans and the coastal waters around islands are among the most vulnerable. Although the transition from sustainable to unsustainable harvesting began with human colonization, it accelerated with colonial activities and increased exponentially between 1950 and 2000. We provide three examples from our nine island groups: Tonga, where subsistence fishing is still an important activity; and Iceland and Japan, where commercial fishing is a central part of the culture. Fish consumption in each of these three countries is three to four times greater than the world average of 16 kg per person per year.

Small, relatively remote island groups such as Tonga have long relied on subsistence fishing, using spears, hand-lines, gill nets, and fences to fish the reefs, lagoons, and tidal flats (Plate 11). These sources supplied most local needs until the 1960s when a combination of increased human population and decreased fish populations led to an increase in off-shore fishing. Subsistence whaling ended in 1978 but whale watching is a rapidly growing tourist industry. Also at this time, an increased availability of air cargo space permitted the development of an export industry of tuna and snapper, largely to Hawai'i. Development of a local cannery in Tonga has been considered. In the 1990s, a limited aquarium trade and commercial sport fishing developed in Tonga but export bans on sea cucumber, giant clams, lobsters, and black coral were instituted to protect diminishing populations.

For Icelanders, fishing is a major industry (Fig. 7.3) and one that the country has vigorously defended. Between 1952 and 1975, Iceland extended its exclusion zone from 4 to 12 to 50 and eventually to 200 nautical miles (370 km) offshore, leading most nations in this process. Iceland's claim on a small, uninhabited island 100 km north of the main island is another way of extending its fishing rights. This extension of fishing rights came as global fishing pressure increased and was resisted by the British in the so-called Cod Wars in the 1970s. The two countries clashed when British trawlers insisted on fishing within the exclusion zone that Iceland claimed. The Icelandic Coast Guard then cut their nets and British Navy warships defended their trawlers. Britain agreed to the fishing restrictions in 1976 after the Icelandic government threatened to close a NATO base. However, fishing resources have been over-exploited by Iceland and its neighbors, leading to the collapse of the herring fishery in the late 1960s. In the 1970s, the cod fishery on the Grand Banks off Canada's coast had largely collapsed and when Canada enforced a 200 nautical mile exclusion zone, it restricted cod fishing. Iceland, along with Japan, still hunts whales, despite international resistance.

Fig. 7.3   Icelandic fishing boats.

The annual fish catch of Japan, about 8% of the world total, rivals that of China and the USA. However, meeting local demand requires that the Japanese also import more fish and fish products than any other country in the world. This level of demand clearly makes Japan a global influence on world fisheries. In the 1970s, when the USA and eighty other countries decided to establish 200 nautical mile exclusion limits along their coastlines, Japanese fishermen were in trouble – as 70% of Japan's annual catch is in what are now considered territorial waters of the USA. Therefore, Japan pays fees and has international agreements with countries such as the USA. Japan is also actively involved in aquaculture, using advanced techniques of insemination and hatching to produce fish and shellfish for domestic consumption. Japan agreed in 1987 to stop commercial whaling, but still kills minke whales and toothed whales (mostly sperm whales) each year for pur-ported research purposes. With increased technological advances in fishing, a planet hungry for protein, and the collapse of many fish pop-ulations, fishing has become of great international significance. Island economies are clearly vulnerable to quotas, exclusion zones, and other international management decisions affecting fishing rights.

7.3.2 Agriculture

Agriculture (including such parallel activities as silviculture and aqua-culture) has substantially altered human–environment relationships.

Early island communities were almost entirely based on subsistence agriculture and the successful manipulation of the environment to grow plants for food was essential for survival. One notable exception was the Guanche people on the Canary Islands, who were mainly goat and sheep herders. Successful manipulations included taro cultivation throughout the tropical Pacific by the Polynesians and the establishment in the 1840s of sheep farming in the native grasslands of the South Island of New Zealand. However, as we have seen in Chapter 6, successfully transplanted agriculture came at the expense of New Zealand's native plants and animals. Failures include attempts by some Scotsmen to farm sheep and cattle on Campbell Island, 700 km south of the South Island of New Zealand. This effort lasted from 1896 to 1931 but failed as a result of isolation, economic woes of the 1930s, and a cloudy, wet climate with rain 325 days each year. Subsequent removal of cattle (1984), sheep (1992), and rats (2001) has allowed native birds to recolonize Campbell Island.

Technological changes during the latter half of the twentieth century changed the environmental impacts of agriculture. Genetic engineering and chemical fertilizers increased crop yields. Technological improvements in dam building, irrigation, tilling, and harvesting allowed for an unparalleled intensification of agriculture. In the British Isles, mechanized farming led to larger fields, more monocultures, destruction of medieval hedgerows and subsequent declines in song bird populations. In Hawai'i, large pineapple farms (Lāna'i, Maui; O'ahu, see Fig. 6.19) or beef cattle ranches (Moloka'i, Island of Hawai'i) created monocultures across large swaths of land. Monoculture farming for exports either continued historical trends on some islands (see Chapter 6) but became more mechanized (New Zealand wool, Icelandic lamb, Canary Island wines), or trends were started on islands where monoculture cropping for export had been uncommon previously (squash production in Tonga for export to Japan). International subsidies for products such as sugar and coffee helped drive the mechanization of agriculture on islands such as Jamaica and Puerto Rico. Yet mechanization of farms meant the end of widespread, small-scale farming; consolidation of farming in the hands of fewer people also led to rural unemployment and migration to cities. We illustrate these changes in agriculture and its associated environmental impacts during the second half of the twentieth century with contrasting examples from four island groups: Japan, New Zealand, Tonga, and Jamaica. We chose these four island groups because of the contrast of wealthy countries, where agriculture as a percent of GDP (Fig. 7.4)

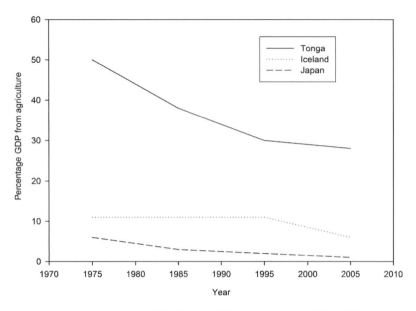

Fig. 7.4   Percentage of GDP derived from agriculture, 1975–2005.
Source: http://earthtrends.wri.org, accessed 22 October 2009.

has decreased to minimal levels (Japan, but also Puerto Rico and the British Isles), with developing countries, where traditional agriculture is still prevalent (Tonga). Jamaica and New Zealand resemble Iceland (shown in Fig. 7.4) where agriculture has remained at around 10% of GDP until recently.

In Japan, land reform following World War II completely altered Japanese agriculture that had been based on rice cultivation for 2400 years. An increase in agricultural production between 1940 and 1960 was caused by the redistribution of agricultural land from larger land-owners to landless peasants, increased use of petroleum-based products such as fertilizers and pesticides, increased mechanization of agriculture, and rural electrification. Wheat and meat consumption increased, human labor was replaced by machinery, and laborers were required to produce cash crops of exotic vegetables and flowers for urban markets. At the same time, a drive to achieve self-sufficiency in rice production for Japan's increasing population came at the cost of land availability for other traditional crops, some of which, such as millet, have all but disappeared from the post-war Japanese diet; the once widespread mixed cropping of vegetables is now much reduced.

Between 1950 and 2000, half the arable land in Japan was converted for industrial and domestic development. There was a breakdown of traditional, cooperative efforts that were needed to maintain rural rice fields and terraces that stabilize slopes. The results included slope erosion, downstream silting, and damage to coastal fisheries, as had occurred during periods of agricultural expansion in the past (see Chapter 6). Fortunately, food quality is again becoming a concern among relatively wealthy, urban Japanese who buy directly from organic farmers. Similarly, coastal farmers, whose crops of seaweed are being destroyed by agricultural runoff, are promoting upland soil restoration.

Agricultural expansion in New Zealand between 1840 and 1940 (primarily clearing of land for grazing by sheep and cattle) was extremely rapid, resulting in conversion of 51% of the country (see Chapter 6). One problem, however, was the phosphorus deficiencies found in New Zealand soils. Beginning in 1919, New Zealand (along with Australia and Britain) mined rock phosphate from Nauru, Ocean Islands, and other tropical islands. These activities damaged the ecology of these islands by removing vegetation and soil and increasing the frequency of droughts, but provided enormous quantities of fertilizer to support New Zealand agriculture. The availability of planes following World War II facilitated the aerial spreading of this "super phosphate," further accelerating agricultural expansion. Another problem with volcanic soils on the North Island of New Zealand was a deficiency of the mineral cobalt. Addition of this critical nutrient to animal feed enabled expansion of agriculture to these soils.

The environmental consequences of agricultural expansion in New Zealand led to classification of ten percent of the land as "severely eroded." Infertile soils, often in areas of high rainfall, were too unstable to support pasture vegetation. By the mid 1970s, concerns over the wholesale alteration of the environment were being seriously considered (echoing earlier but timid voices from the 1940s). Pasture grasses were not stopping erosion. Waterways were being polluted by animal waste equivalent to waste production from a human population of 150 million people (New Zealand in the 1970s had only 3 million people). Recreation in the few remaining native landscapes was being encouraged and efforts were begun to protect, restore, and expand these areas. Gorse, a spiny shrub that invades pastures and is the scourge of New Zealand farmers, is being used (in one valley) to help restore the soil by fixing nitrogen and acting as a nurse plant for native trees. As New Zealanders realized how they had damaged the

Fig. 7.5   Radiata pine plantations that have been recently clearcut, Nelson, South Island, New Zealand.

land, they began to re-evaluate their dedication to applying English-style farming to their unstable, often nutrient-poor soils. However, changes in land use did not necessarily mean improved environmental conditions for native species. Radiata (Monterey) pine plantations began to replace pastures as this tree, rare in its native California, became the foundation for New Zealand's modern forestry industry. The switch from sheep pastures to pine plantations produces habitat better suited for some native plant and animal species and provides better protection of slopes against erosion, but benefits are often undone when plantations are clear-cut for timber production (Fig. 7.5). In fact, the process is now being reversed in many parts of New Zealand, as pine plantations are being converted again to pasture, this time for the more lucrative dairy industry. New markets for dairy products such as India and China are driving an expansion of dairy farming into areas where previously sheep, in low densities, grazed semi-natural grasslands on the South Island. These new dairy farms replace the grasslands with monocultures of introduced grasses and need extensive irrigation because they are located in dry climates. The environmental effects of water extraction from nearby rivers for irrigation, fertilization to improve pasture growth, and effluent from cattle contaminating ground water are emergent problems caused by the current boom in dairy farming in New Zealand.

Tonga remains today a nation still largely dependent on agriculture, both for subsistence of the local population and, increasingly, for cash crops that are sold both domestically and abroad. About two-thirds of exports are still agricultural products. During the twentieth century, Tonga's main export was coconut products, especially copra (dried coconut meat from which coconut oil is extracted); a market that had all but collapsed by the 1990s. In the 1980s, a market for squash developed because of high prices that Japan was willing to pay. However, this market is unstable and production has fluctuated due to competition from Vanuatu and Mexico, disease, poor weather, and overproduction. Vanilla beans, already an export crop from Tonga in the nineteenth century (see Chapter 6 and Fig. 6.23), again flourished as an export in the early 1990s. Recently, vanilla bean exports have suffered from competition from Central America and Madagascar. Production of kava, a mild and legal narcotic introduced to Tonga at the time of Polynesian settlement, has increased in recent years and is now exported to communities of expatriate Tongans. Products grown largely for domestic markets and subsistence include traditional Polynesian crops such as bananas, yams, taro, and sweet potatoes, as well as more recent introductions such as cassava and oranges (see Fig. 6.22). Population growth and particularly a gradual shift to cash crops from subsistence agriculture has led to expansion of agricultural land (already occupying about 70% of Tonga), increased use of chemical fertilizers and pesticides, and less agricultural land that lies fallow. Pesticide residues from intensive farming for crops such as squash now contaminate ground water reservoirs on Tongatapu, the most densely populated island. Although there is still an emphasis on increasing productivity and little concern about soil and water quality, there is a nascent organic farming movement, mostly for vanilla. Transportation from croplands to markets is a large hurdle because of the remoteness of many of the islands, so subsistence on locally grown products is still a logical choice for many, at least until demands from international markets stimulate further development of cash crops.

Jamaica's colonial history left an agricultural legacy of crops produced for export, such as bananas and sugar (see Chapter 6). A series of land reforms during the last 100 years has encouraged the expansion of small farms to marginally steep and infertile lands, resulting in substantial loss of forest cover and soil erosion. Soil productivity may have declined as much as 30% as a result of erosion. Slash-and-burn methods of farming, also conducive to erosion, are still used in mountainous terrain (Box 7.1). Simultaneously, consolidation of sugar, dairy, and other major agricultural industries in lowland farms has led to centralized and more mechanized agriculture, beginning with the introduction of irrigation in 1868. The larger farms have reduced the patchiness of the landscape provided by many small farms, largely to the detriment of wildlife.

Box 7.1. Fire in the mountains

The Cinchona Botanic Gardens in Jamaica sit on a spur of the Blue Mountains at 1400 m at the end of steep, rutted earth roads that are treacherous after rain, and are often blocked by landslides. The Gardens were a field station for me when I was conducting research in the rainforests, a walk of an hour or more from the Gardens. The Gardens are perched above steep slopes of the Yallahs River Valley, and below the gardens these slopes are almost entirely denuded of the original forest cover. These slopes were first cleared more than 200 years ago to make way for coffee plantations that relied on slave labor. After the coffee boom ended and after emancipation in 1838, freed slaves came up to the mountains to start new lives, and feed their families by farming these steep slopes. Their descendents are the farmers who live in the Blue Mountains now, living in small towns like Westphalia and Hall's Delight, and walking each day, often for an hour or more, to their small farms on the slopes in the district. The soils are thin, and fractured bedrock is strewn through the soil. It is hard to stand on some of these slopes without slipping downhill. Small logs are used to help stabilize parts of the slopes, and crops are planted here. Thyme and spring onions are common crops, as are turnips – this is quite a temperate climate high in the mountains of this tropical island. How the crops fare, on the leeward side of the Blue Mountains, is entirely dependent on rains. Too little, and the crops will not grow. Too much, and the hillsides will erode away, taking the crops, and sometimes entire farms, with them. I knew a family nearby who lost both their home and their land during the rains of Hurricane Gilbert. Most of the crops are grown to support a farmer's family.

Box 7.1. (cont.)

The work is hard, and many children are taken out of school to assist their parents when it is time to plant or reap crops, so that their families can eat. Any surplus that is grown will be taken to market, as long as the roads are open, and as long as a reliable vehicle is working, with as many people and produce crammed inside as possible for the zig-zag route down the mountain.

During the dry months the mountains burn. The parts of the slopes that are no longer farmed, left fallow after the thin soils are exhausted, do not return to forest but instead are invaded by molasses grass, native to Africa and highly combustible. Fires are set to clear land to re-establish farms (see Fig. 3.18), but many are accidental, so by the end of the dry season, many areas are left as blackened, burned grass with a few thin woody stems, bare soil, and bedrock outcrops. A fire set low on the slopes, accidental or deliberate, quickly races hundreds of meters up slope. Being at the top of a hillside when fires are burning below can be very unnerving. At the head of the Yallahs Valley is the little amount of rainforest that survives on the leeward slopes of the Blue Mountains. As the soils further away are exhausted and no longer support crops, farmers in the district walk even further to cut small patches of the rainforest, leave the fallen stems to dry during the dry season, then burning them to create an ash bed that they will farm. In this way, and with occasional resurgences in coffee plantations, the remaining natural forests are threatened. Re-creating forest cover to stabilize the slopes, assisting farmers to develop more sustainable methods of agriculture in the Blue Mountains, and overcoming rural poverty have been important management goals in the valley for at least the last 50 years. PJB

Since World War II, agriculture in Jamaica has declined to about 6% of GDP, a value less than in most developing countries. This decline can be attributed to migration of farmers to urban areas or abroad and development of bauxite mines on farm land since 1950. Farmers attempted to stem the decline by diversifying crop production to include citrus, cocoa, coconuts, coffee, ginger, and tobacco. However, independence from Britain in 1962 and the entry of Britain into the European Union in 1972 spelled the end of preferential markets and sent Jamaican agriculture into further decline. Jamaica still exports some sugar and it also exports rum, which is distilled from sugar. Marijuana is also a valuable export crop, contributing substantially to some rural incomes, albeit unofficially. Jamaica has long been relatively self-sufficient in fresh vegetables but has not achieved self-sufficiency in other staples such as rice, beans, dairy, fish, timber, and other products. The Ministry of Agriculture is allotted less than 1% of GDP but hopes to promote food security, productivity, infrastructure, and marketing for agriculture. Meanwhile, GDP per capita (USD 3622 in 2005) remains lower than any of the island groups except Tonga (USD 2169) and about 20% of Jamaicans are below the poverty line. The environmental consequences of poverty are mixed, because slash-and-burn techniques and multiple subsistence crops can have detrimental local effects while agribusiness impacts the landscape at larger scales. The disintegration of rural agriculture in Jamaica is illustrated by the loss of up to 25% of some crops by theft. Crop theft can occur because poor families may steal to stave off hunger, but can also be more systematically undertaken to sell the stolen crops in markets. Local productivity has been further hampered by discount sales of large quantities of crops such as potatoes from Idaho (USA) and dairy products from Wisconsin (USA). While the role of traditional agriculture is diminishing in Jamaican society, it has not yet been abandoned.

As illustrated by these four examples, the mechanization of agriculture generally increases erosion and involves increased use of pesticides and chemical fertilizers that pollute surrounding waterways. Modern agriculture also directly pollutes the air through emissions of nitrous oxides and methane. Centralization of farm ownership contributed to farmers migrating to cities. Both of these issues are not unique to islands, but are particularly tractable to study on islands because of their finite and isolated geography. When farms are abandoned, they can revert to secondary forests, as occurred in Puerto Rico in the 1950s. Another impact of the abandonment of farms has been the invasion of weeds. On the Island of Hawai'i, for example, molasses grass and other flammable plants have increased fire frequency in ecosystems not accustomed to being burned (see Chapter 6). Agriculture has created human's oldest and most enduring impact on the land and yet introduction of chemicals, secondary forest invasion by non-native species, and increased fire regimes suggest that novel changes are still occurring.

Different island cultures have addressed the modernization of agriculture in unique ways, from rapid abandonment of traditional techniques (Japan), to rapid expansion of agriculture for export markets (New Zealand), to tentative steps toward modernization with a foothold still in the subsistence mode (Tonga), to a mix of traditional and modern modes (Jamaica). In New Zealand, sheep farms are being replaced by radiata pine, dairy, and now wine production, each with its attendant environmental costs. In Tonga, squash exports flourished and then crashed in synchrony with Japan's economy, but pesticide pollution from squash production remains a problem. Some of the money for squash was spent on importing second-hand Japanese cars, increasing the impacts of roads, traffic, and pollution in Tonga. Now disposal of old cars is an additional problem. Jamaica's sugar and banana markets crashed when global trade agreements increased competition from other producers. On other islands, sugar cane has been replaced by suburbs (Puerto Rico) and macadamia nut production (Hawai'i). There have been severe environmental consequences of completely replacing traditional, local-based agriculture with cash crops intended for unreliable world markets. Indications that consumers in some areas are now demanding quality, locally grown products is a good sign for the redevelopment of food security and market stability through self sufficiency for island economies. Consumption of locally grown produce may also be more environmentally friendly because it entails less fuel for transportation and less use of insecticides and pesticides on farms that are typically kept small.

7.3.3 Extractive industries

Historically, human societies have been built on the successful extraction of resources from the sea (fishing) and the soil (agriculture). Some islands have historically provided other types of resources as well, such as dense basalt (Tonga, Hawai'i) and obsidian (Hawai'i, Japan) for making adzes, spear tips, and knives, or greenstone (New Zealand) for making jewelry. Early European colonization led to extraction of gold from Puerto Rico and gold, silver, and copper from New Zealand. More recently, extraction techniques have improved and trade has developed other products to satisfy both local and global demand for mineral products, clean water, and energy. For example, although Jamaica is the world's fourth largest producer of bauxite (see below), it also exports gypsum, lime, marble, and gold. Diatomaceous earth, which is used in filters and as an absorbent, and is a component in dynamite, is dredged from Lake Myvatn in northern Iceland. Some island resources have yet to be utilized such as copper in Puerto Rico or copper, gold, silver, sulfur, and zinc from the sea floor near Tonga. Aquatic resources are also extracted, either directly in the form of clean water or indirectly, by tapping into the energy supplied by running water. Iceland has several aluminum smelters that use hydropower. Progress has been made toward regulation of extractive industries but they remain major contributors to habitat alteration and degradation. In this section, we explore the environmental consequences of extractive industries and note how technology has increased the pace of such extractions. We use four examples, coal mining in the British Isles, bauxite mining in Jamaica, artesian aquifers in Puerto Rico, and hydropower in New Zealand.

In the British Isles, coal mining occurs primarily in England, Wales and southern Scotland and has been a mainstay of the local economy for several centuries, peaking in the early 1900s with over 250 million tons extracted per year (compared to current production of 30 million tons). Nevertheless, current high energy demands mean that the British Isles import 70% of the coal used (Plate 16). Economic downturns and depletion of coal supplies have meant the closure of most coal mines in England (down from 170 in 1984 to 10 today). These closures have been resisted fiercely by coal miners. The industry was nationalized in 1947 and then privatized again in 1994. Old mines are being reclaimed as nature parks, business parks, and even energy-efficient housing communities. However, in the last decade an expansion of open pit coal mines has occurred, as a result of increased costs of other fossil fuels. Open pit coal mines have many environmental impacts. The mines remove the plants, animals, and soils on the surface, are notorious emitters of carbon dioxide, and create harmful dust. Coal mines also impact local water supplies, especially when flooding moves mine wastes and causes contamination of nearby landscapes. Occasionally, land subsides around mine shaft openings and damages buildings. Reclamation of mined areas is done through filling of mine shafts and pits, replacement or importation of topsoil, forest plantings, and wetland reconstruction.

Bauxite, a critical component of aluminum, has been a major component of Jamaica's economy since 1952, currently representing three-quarters of all legal exports. Bauxite is found in about 25% of the surface rocks of Jamaica. Of particular concern are those reserves found in the central Cockpit Country or karst hills. If these reserves were removed, at least 65 endemic plant species and four endemic animal species could be threatened. The forest cover in the karst hills provides an important function as a water catchment and flood control for people living on Jamaica's coastal plains. While there are

Fig. 7.6   Bauxite mine near Porus, Manchester, Jamaica.

other environmental pressures on the Cockpit Country (notably heavy use of tree saplings to support growing yams), bauxite mining is the most destructive. Bauxite ore is extracted after the topsoil has been removed. As with coal mining, environmental problems include not only the total destruction of the plants, animals, and soil (Fig. 7.6), but also the production of toxic wastes. For every ton of bauxite that is removed, a ton of alkaline mud remains. This mud was initially stored in mined-out areas and diked valleys, but water contamination by sodium hydroxide led to unsatisfactory efforts to use sealed, artificial ponds. Other environmental damages include the loss of arable land to extraction, waste deposition, transportation networks, and coastal refineries. Air pollution, dust, and spills at storage facilities and along coastal reefs are also a problem. The companies that mine Jamaican bauxite (all are foreign owned) have had some success in recovering mined lands and establishing farmland or plantation forests, but the original ecosystems and topographical complexity are not recovered. Conservationists argue that tourism (comprising at least one fifth of Jamaica's economy) is more sustainable than surface mining and should be a future priority. The global recession in 2008–2010 has slowed bauxite production, a benefit for the environment of Jamaica, if not its economy.

Puerto Rico has some of the purest water in the world in its underground aquifers because the water has been filtered over millennia through extensive limestone deposits. These artesian aquifers provide a resource essential to the manufacturing of chemicals.

Beginning in 1950, US-based pharmaceutical companies began to establish factories in towns such as Barceloneta and Toa Baja on the north shore to take advantage of the easily accessible, high-quality water. Encouraged by generous tax breaks from the US government, by 1978 there were 57 pharmaceutical companies responsible for 32% of Puerto Rico's GDP. Large percentages of the global profits of these companies have come from products manufactured in Puerto Rico. However, with lax environmental regulations, aquifer water levels have been drawn down tens of meters; pumping is now required to obtain the water. In addition, waste water has been dumped in sink holes and pumped back into the deep aquifers. Wastes have also been dumped into local rivers or offshore, damaging fisheries and infiltrating ground water. One dramatic event occurred in 1981 when a depression filled with waste water from a pharmaceutical company collapsed following a particularly heavy rainstorm and 1.2 million gallons of polluted water disappeared in less than a minute into four new sinkholes. Pharmaceutical companies have depleted the fresh water to the point that brackish water is now entering the aquifers. Recent efforts have centered on maintaining clean water for local residents, detecting contaminants, and educating people about how to reduce further ground water pollution.

Hydropower contributes 70% of the energy used by New Zealand but development of this resource has led to environmental changes. A frequent concern is the effects that changing water levels from hydropower generation have on wildlife. For example, streams flowing down from the volcanoes in the central part of the North Island have largely been channeled into Lake Taupo (increasing its input by 20%) but one source (Tongariro River) was left undisturbed to accommodate breeding grounds of the introduced brown and rainbow trout. Two power plants on Lake Taupo and ten more on nearby Waikato River provide 13% of New Zealand's electricity. One of those plants is a gas- and coal-fired power plant for which the river provides cooling water. In order to protect the 19 native and ten introduced fish and other aquatic organisms in the river, the plant managers are not allowed to warm the river water to over 25°C. On the South Island, hydropower schemes have altered water levels and flow patterns on braided river channels in the MacKenzie Basin, reducing populations of the black stilt, one of the rarest wading birds in the world (with less than 50 adults both captive and wild). In 1959, a proposed 30 m increase in the level of Lake Manapouri to augment hydropower production triggered a 13-year protest (their motto: Damn the Dam!) that captured

Fig. 7.7   Dam construction, Yakushima, Japan.

the attention of the growing global environmental movement. The proposed power increase was designated for the sole purpose of smelting bauxite imported from Australia into aluminum for export. The protest was fueled by concerns about impacts on wildlife and the preservation of the natural beauty of the area; a petition was signed by 10% of the New Zealand population. The protest was eventually successful, and power generation causes lake levels to fluctuate only within acceptable limits. A citizen group of guardians now monitors the levels of Lakes Manapouri, Te Anau, and Monowai. Environmental damage by dams is not isolated to New Zealand (Fig. 7.7). Dams are environmentally destructive on other islands. Yet similar protests against hydropower projects on other islands (notably the damming of the Kárahnjúkar region in eastern Iceland) have not been successful. Given the impacts of hydropower (alteration of braided river habitats, lake level alterations, and silt build up behind dams), New Zealand is gradually diversifying with other energy sources including geothermal, solar, and wind. Tidal power is also under consideration.

However, the high demand for hydropower currently threatens other watersheds in New Zealand, such as the Mokihinui River, previously designated for conservation.

The threats posed by mining and other resource extraction for industry and energy production are not going away, despite their environmental impacts. The demand for raw materials and energy to fuel our modern economy keeps increasing due to population growth, improvement of standards of living (and therefore consumption patterns) of developing countries, and development and adoption of new technologies such as cell phones that use rare minerals. Some progress has been made to educate islanders about the links between consumer products and their environmental impacts.

7.3.4 Military activities

Islands are a frequent target for both military aggression and defensive activities because of their perceived strategic value. Most of the nine island groups have either been bombed or served as military outposts in the last 70 years, from the bombings of Pearl Harbor (Hawai'i); London, Liverpool, and Coventry (England); and Tokyo, Hiroshima, and Nagasaki (Japan) to the build up of (mostly American) bases on many of the island groups. Environmental consequences of such military activities have increased as military technology has improved but the consequences are more nuanced than is often realized. While most military activities are harmful to the environment, some are benign or inadvertently beneficial. We will discuss each of these aspects of military activities in this section and note how technological advances have increased environmental consequences.

Immediate environmental damage from warfare or military exercises includes chemical wastes, severe soil compaction, fire damage to fields and forests, and pollution by depleted uranium from armor-piercing incendiary shells. In Japan during World War II, about 15% of forests were felled for the war effort, including many old-growth forests. Forests of pine trees were uprooted to obtain a distillate to make motor oil from the roots – an effort that, ironically, never worked. Many songbirds were killed in Japan both for food and for feathers to make pillows and quilts for soldiers. Military exercises damage land and water resources, especially through practice bombing and compaction from tank tracks. The then-uninhabited island of Kaho'olawe near Maui in Hawai'i was used for bombing practice and testing of new military technology by the US military from 1920 (officially from

1941) until 1981. In addition to disrupting the native plant communities, exacerbating soil erosion, riddling the land with craters, and leaving behind unexploded ordnance, bombing apparently damaged the island's fresh water aquifer from a large blast that had the equivalent force of 500 tons of TNT. The military returned control of the island to the State of Hawai'i in 1994.

Longer-term environmental consequences from military activities include residual battlefield scars and impacts on natural resources. The wholesale cutting of forests in Japan led to an increase in erosion, flooding, and pine bark beetles that feed on dead or damaged trees. Even the invention of DDT as an insecticide was a result of military activities because western scientists invented it in order to replace pyrethrum that had been produced almost exclusively from Japanese chrysanthemums no longer available during World War II. DDT use has been associated with myriad environmental issues around the world including its negative effects on the survival of bird embryos.

Military construction is another impact with long-term environmental consequences. After World War II, the US military built permanent bases in Puerto Rico, Hawai'i, Iceland, the British Isles, and Japan. Military bases often occupy large tracts of land, such as the 3200 ha Roosevelt Roads and 10 800 ha Vieques naval stations in Puerto Rico, the 23 700 ha Kadena base in Okinawa, Japan, and the 44 000 ha Pohakuloa base on the Island of Hawai'i. Maintenance, expansion, and utilization of these bases have on-going environmental costs, including initial loss of biodiversity, compaction of soils and disruption of watersheds, littering with unexploded ordnance, radioactive pollution, leaked fuel, toxic battery fluids, lead contamination from ammunition, and noise pollution from low-flying aircraft.

Military actions can sometimes have surprisingly benign effects on the environment, such as providing habitat for some species. Insects, fish, mammals, and plants recovered very quickly in Tokyo after the extensive bombing in World War II and both wild plants and urban gardens actually grew prolifically until urbanization again dominated and reduced light, water, and soil available for plant growth. By the mid 1950s, most environmental scars of the war were gone from Japanese cities. Military activities can be inadvertently beneficial when they decrease exploitation of a natural resource. For example, deep-sea fishing was halted during and after World War II. Resource protection is another positive effect of military activities. Military bases limit access and development and therefore often become critical habitat for wildlife, including endangered species. For example, overfishing and silt runoff from farms have severely damaged coral reef communities around Puerto Rico. Yet within the two large bays encompassed by Roosevelt Roads Naval Station in eastern Puerto Rico, reef communities remained remarkably healthy, at least until the base was closed in 2004. The former naval base is now being converted into an airport and other developments.

The Pohakuloa military base, established in the 1940s on the Island of Hawai'i, is the largest military facility in Hawai'i and it has also had some beneficial effects on the environment. It contains habitat for a number of endangered plants and animals as well as several archeological sites. In 2001, a 15 600 ha area was fenced to exclude feral goats and sheep that were eating the rare plants. In addition, efforts are being made to re-establish native forests upon which the endangered palila bird depends.

The dismantling of military bases can trigger environmental concerns. Culebra and Vieques, two small islands off the eastern Puerto Rican shore, were sites of military activities beginning in the 1940s. Despite a civilian population of several thousand, Culebra received the brunt of practice bombing and missile firings, which were conducted as military training exercises and peaked in the late 1960s. Citizen protests finally forced the US Navy to leave Culebra in 1975 but they then increased exercises on 70% (10 800 ha) of nearby Vieques. To the great relief of Vieques residents, the US military left in 2003. Fears of resort development were unfounded because today the military lands are a national wildlife refuge. It is ironic that tourism is considered as much a threat to an island as bombing had been. Despite the occasional positive aspects of military bases providing refugia for endangered plants and animals, the net impact of military activities is gradual environmental degradation, whether from warfare or preparation for it. Sometimes military disturbances increase the rate of invasion of non-native species by reducing competition from native plants and opening up new habitats or by moving non-native species among islands. Technological advances have increased the negative impact of military activities. For example, World War II began with biplanes and Sherman tanks and ended with jet-propelled fighter planes. New models of tanks and planes keep appearing, each with an additional impact on the environment, directly through soil compaction, paved tank routes, and airports, and indirectly through fuel consumption and greenhouse gas emissions. Long-term radiation effects from nuclear weapons negatively impacted human health in Japan (Hiroshima, Nagasaki) and on other Pacific islands such as Bikini and

Rongelap in the Marshall Islands. Preparing for war and waging war are hard on the environment.

7.4 POPULATION GROWTH

7.4.1 Urbanization

Historically, civilizations rarely had less than 10% of their population in urban areas. Extensive urbanization is a very recent phenomenon. As Carl Sagan described it, if the creation of the universe occurred on 1 January, humans appeared on 31 December, the first permanent settlements in the last minute of that last day of the year (10 000 years ago) and extensive urbanization only in the last second. With the advent of capitalism and the Industrial Revolution, cities grew in number and size and became areas that concentrated the production and sale of goods. Simultaneously, there were declines in mortality and increases in population. The rapid concentration of people into cities first occurred in northwestern Europe, especially the British Isles. This rapid rise in European urban populations, which began in the latter part of the eighteenth century, slowed after several generations, when about 75% of the population lived in urban centers. Other nations followed similar patterns of urbanization such as in the USA where cities grew particularly rapidly from 1880 to 1920. Between 1950 and 2000, the growth of cities again accelerated across most developed nations as people moved from rural to urban areas for employment and immigrated from other countries. Between 1950 and 2000, the global human population increased from 2.5 to 6.1 billion, in part due to advances in medical and agricultural technology that resulted in humans living longer and having more food to eat. However, among the nine island groups, the increase in the degree of urbanization has varied greatly (Fig. 7.8). The most rapid increase in urbanization occurred in Puerto Rico, which went from 41% of its population living in urban centers in 1950 to 95% in 2000. Puerto Rico is now also the most urbanized of the island groups, with 99% of its population living in urban areas. Jamaica's urban population changed from 24 to 52% of the total population, not quite as big a change as in Puerto Rico, but still a substantial climb (Fig. 7.9). The increase in urban populations in Jamaica resembled world-wide averages during the same time period. Tonga, by far the least urban of the island groups, changed least (from 13% urban in 1950 to 23% urban in 2000). Among the remaining island groups, Iceland and the UK (British Isles minus the Republic of Ireland)

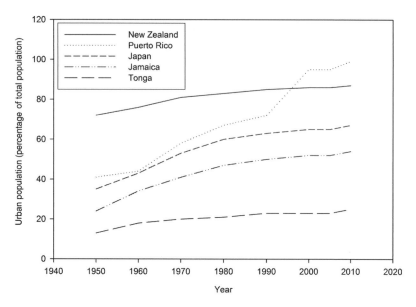

Fig. 7.8   Percentage of population considered urban (as defined by each country). *Source:* http://esa/un.org/unup (accessed 27 October 2009).

Fig. 7.9   Abandoned house at Nonsuch, Portland, Jamaica.

resemble New Zealand (shown in Fig. 7.8) and all three are substantially more urban than Japan. The Republic of Ireland remains as rural as Japan and Jamaica. In this section, we highlight recent urbanization in the British Isles, the Canary Islands, and Puerto Rico, where the triggers for urbanization were the Industrial Revolution, tourism, and land reform, respectively.

Fig. 7.10   Dense housing blocks in Salford, Manchester, England.

The British Isles led the world and the other island groups in the process of modern urbanization. By its first census in 1801, one-third of the British population lived in cities. This number expanded to about 50% by 1857 and 78% by 1901. London was the first city in the world to record over 5 million inhabitants (in the 1891 census). By 1950, there were eight cities in the world with 5 million people (83 with 1 million); today there are 46 cities with 5 million people (and almost 500 cities with 1 million people). Yet urban sprawl has been contained in the British Isles because of the high density of dwellings within cities (Fig. 7.10), the ability to commute to cities using public transport, and restrictions that limit growth at the city edges. However, the attraction of single family homes and larger lots with yards has seen an upsurge in the British Isles and elsewhere in Europe in the last few decades, with attendant impacts on landscapes and natural resources including water. The Republic of Ireland, however, is on a different trajectory to that of the UK. Ireland has maintained more of its rural character, but still had more than 50% of its population living in cities by the year 2000.

Manchester was a central point for urbanization in England and provides a good example of how a city responds to changing environmental challenges. In the middle of the nineteenth century, Manchester's population grew rapidly as the city became a global center for textile manufacturing (Fig. 7.11). Cotton was imported from the southeastern USA and textile products were exported to global markets. Since the 1960s, Manchester's population has been stable at 2.5 million. Water and air pollution accompanied the industrialization

Fig. 7.11   Manchester Shipping Canal and Salford Quays, Manchester, England.

of Manchester. Rivers served as disposal sites for sewage and only in the 1990s did they become clean enough to eat the fish removed from them, in part because of the collapse of the industrial base of the city. Air pollution has also been an on-going problem in Manchester, as in other urban centers, but the causes have changed over time. When coal dominated as a fuel source, clean air was seen as an unrealistic, utopian ideal. However, in the 1940s, Manchester created smokeless zones, sulfur dioxide decreased, and human health conditions improved dramatically. Natural gas from the North Sea became available just as national Clean Air Acts were implemented in 1956 and 1968, but tall chimneys moved pollutants into higher air streams with resultant acidification of lakes in Scandinavia. The most recent form of air pollution comes from oxides of nitrogen in traffic emissions, a challenge met in part by renovating city trams. Manchester has recently been a leader in attempting to provide a livable environment in an area of high population density. Today the city is reclaiming landfills, rebuilding aging infrastructure, and taking responsibility for waste that it formerly exported to other ecosystems.

The Canary Islands are highly urbanized, largely because of the concentration of tourism. Urbanization brings increased environmental pressures by tourists on natural areas such as beaches and mountains, as well as increased production of waste. Tourism has largely replaced a series of agricultural monocultures that were exported until the 1920s (see Chapter 6 and Section 7.5.1). Therefore, hillsides that were once carefully terraced for agriculture have been abandoned and are in disrepair, leading to soil erosion. Local agricultural production supplies some food to the tourist trade, but most food is still imported as packaged products, increasing the pressure on land-fills and contaminating ground water with plastics and the chemicals they contain. Recycling is improving on some islands (Tenerife and Gran Canaria) but still minimal on others (La Gomera) because of a lack of infrastructure for collecting and processing (or transporting) such wastes. Water resources are also hampered by a lack of recycling, losses of up to 50% during transportation due to leaky pipes, and the pollution and near depletion of aquifers. Some efforts have been made to improve the urban environments in the Canary Islands. Legislation regulating land (1985), water (1990), and tourism (1995) has attempted to protect and integrate the management of natural resources. One effort, spearheaded by the European Union, is to increase natural gas storage capacity and speed development of renewable energy technol-ogy, such as wind turbines, to avoid current dependency on oil, relied on today for 90% of energy needs. Other efforts include increasing fuel efficiency, urban renewal that removes vehicles from city centers, and the addition of parks and other green areas such as was done in the 1990s in Santa Cruz on Tenerife. In order to sustain their current tourism-based economy, Canary Islanders recognize that environmen-tal resources must be protected. For example, they led a successful campaign to secure UNESCO World Heritage status in 1986 for rainfor-ests in Garajonay National Park on La Gomera and in 2007 for the lava landscapes of Teide National Park on Tenerife.

Puerto Rico is also a highly urbanized island, with 439 people per km^2 (the 24th densest population in the world), 11% of land in high-density urban centers and 40% of the island in lower-density urban sprawl. The degree of urbanization increased sharply in the 1950s when farmlands were left fallow. Operation Bootstrap, promoted by the Puerto Rican government in the 1950s, shifted Puerto Rico's economy from agriculture to manufacturing and tourism. Therefore, despite improved living conditions in rural areas (paved roads, elec-tricity) there was a net migration to the cities. During the next four decades, the rural Puerto Rican landscape became dominated by sec-ondary forests that grew in what were once abandoned pastures. By the end of the twentieth century, however, population increases led to urban dwellers returning to their family properties and turning former pastures into homes from which they commuted to the cit-ies. Not all Puerto Ricans that migrated to the cities (primarily to San Juan) enjoyed a better standard of living and many became residents of urban slums that developed in numerous high-rise, low-income

Fig. 7.12   Eastern edge of suburban Kyoto, Japan.

rental apartments. In addition to immigration, Puerto Rican cities grew due to high birth rates that are only today declining to just above replacement levels.

San Juan has one of the worst cases of sprawl of any city in the world. The city extends 35 km inland from the city center. San Juan has a population of 2.2 million living in an area of 2310 km², for a population density of 960 people per km². In comparison, Manchester, England has a similar population size but extends over only 558 km²; Osaka, Kobe, and Kyoto in Japan (Fig. 7.12) combined extend over a similar area as San Juan but have a total population of 15.5 million. The environmental problems of Puerto Rico's burgeoning cities have been tackled irregularly. Some headway was made in the early twentieth century with reforestation of north coast reserves to protect urban watersheds and with mangrove protection in the 1980s. An urgent issue is how to handle the 600 kg of trash generated per person per year in San Juan where the 30 landfill sites are nearing capacity. One suggestion is to begin incineration to reduce waste volume by 90%.

In western Puerto Rico, research is on-going into water quality and sources of nutrient pollution in Mayagüez Bay.

Urbanization leads to the loss of rare plant and animal species, an increase in non-native plants and animals, and an increased movement of plant and animals species across an island group. Yet urban planning, particularly for environmental concerns, has been generally nonexistent. Some exceptions occurred in Puerto Rico and the Canary Islands, both Spanish colonies, where streets and town squares were meticulously designed. Some urban planning also occurred in New Zealand cities, such as the British-planned Christchurch and Dunedin. The development of necessary sewage systems, coping with diseases aggravated by crowding, dirty living conditions, and other uniquely urban issues all stimulated major progress in engineering and public health. Yet modern cities make such a large and devastating impact on the environment that the formal studies of urban ecology and urban planning are still catching up with the rapid changes. Given that most of the world's population now lives in cities, discovering how to make urban living more sustainable is imperative. Any large-scale approach has global consequences because shifts to construction with bamboo, development of renewable energy, and waste recycling in even a few cities will impact world markets. Examples of environmentally conscious cities exist, such as Curitiba, a city of close to 2 million in Brazil that has over 1000 parks and green areas and whose citizens use 23% less fuel per capita than the national average. Island cities such as Honolulu, Hawai'i that import 90% of their energy needs are beginning to take steps to curb their fuel consumption by promoting solar power, electric cars, and public transportation.

Urban sustainability is impossible if traditional models of economic growth prevail. Even if cities were to magically shrink to take up less than their current 2% of the Earth's surface, they still use 75% of the world's resources. These resources come from farmlands, forests, watersheds, and reefs, and their use generates waste that need disposal. Cities would be sustainable only if they did not damage or exceed the capacity of these resources and we know of none that do. All efforts at improving sustainability of cities are part of a larger paradox of unlimited growth from limited resources. A piecemeal approach, such as by adding urban parks or reducing water consumption, ignores the overwhelmingly larger issues of population growth and pollution. Atmospheric pollution is now leading to climate change that, by triggering sea-level rise, will inundate parts of many of the cities in the world. Of course, there are success stories in addressing urban pollution. The Thames River that flows through London, England was considered unfit to drink by 1610. By 1858, it was called the Big Stink and its fish and the birds that fed on them had died. Despite some progress in the late nineteenth century, by 1950 the Thames was again a foul place devoid of fish or oxygen. Greatly improved sewage treatment facilities built between 1964 and 1974 cleaned the river up and wildlife again flourishes. Another success story involves Tokyo, Japan where air quality after World War II became so bad from unfettered growth that traffic police were issued oxygen masks. Fatal red tides and urban subsidence from the pumping of ground water also plagued the people of Tokyo until strict environmental laws were passed in the 1970s. Today, Tokyo is aiming to become the world's most eco-friendly mega-city. These efforts involve recycling of wastes on a human-made island offshore; trash is incinerated to obtain heat to generate electricity while the ash is turned into building materials. Another effort is the construction of green buildings with features such as roof-top rice paddies. Vegetation on a roof reduces the cost of heating and cooling. Other rooftops feature solar panels that are being subsidized by the municipal government. Tokyo is a pioneer among large cities in its efforts to become more sustainable.

7.4.2 Remittance cultures

With improved transportation and better job opportunities in other countries, emigration by native islanders accelerated after World War II, resulting in more Puerto Ricans and Tongans, for example, living abroad than on their home islands. Emigration, however, did not match immigration or *in situ* population growth, so each of the island groups that we consider experienced net growth between 1950 and 2000; net emigration did depopulate other islands during that same interval, such as Niue and some of the Cook Islands in the Pacific Ocean. Despite the large exodus from the home islands, the bond often lasts, resulting in the development of remittance cultures, whereby wage earners send money home to relatives, contributing a large proportion of total island income. This money can move from mainland to island such as contributions sent from New York City to Puerto Rico; between islands such as from the UK to Jamaica, New Zealand to Tonga, or Hawai'i to Japan; and from island to mainland such as from the Canary Islands to Africa. The ecological impact of such a steady infusion of money is unclear but might mean a smaller footprint through less active local agriculture. Most likely, however, the relative wealth islanders receive from remittances is probably used in ways that increase rather than decrease their impacts on the environment. Families receiving remittances, for example, tend to have larger homes and newer vehicles. Remittance cultures therefore provide a subtle but important influence on island environments.

7.5 WEALTH AND LEISURE

The years following World War II were marked by an increase in living standards for many people around the world. Japan and the British Isles rebounded from the physical devastation of the war and, along with New Zealand, from the high mortality rates incurred during the war. Iceland and Puerto Rico began to climb out of poverty while Tonga, the Canary Islands, Jamaica, and Hawai'i increased the production of export products for world markets. New economic activities led to improved financial stability and time for the development of leisure activities. Islands became accessible to visitors with improved transportation (especially air travel) and building of tourist facilities to promote sight-seeing, golf, beach activities, and sport hunting. In this section, we discuss the environmental consequences of the development of a tourism industry and impacts of sport hunting on the nine island groups.

7.5.1 Tourism

The improvement of air travel after World War II helped promote an increase in tourism as previously remote islands became regularly accessible. Tourists, encouraged by airline companies such as Pan American, began to flock to islands that made efforts to present themselves as entertaining destinations. Island tourism has thrived on at least three types of attractions: scenery, recreation, and culture. Scenic delights that are typical island attractions include beaches, rocky coastlines, volcanoes, and coral reefs. Recreational opportunities include water sports, fishing, hunting, and golf. Cultural attractions include unusual and gourmet food, ethnic crafts and festivals, vibrant city life, and pastoral villages. Most tourists come to enjoy the natural and social scenes and then depart. However, some stay and bring new environmental perspectives with them. Yet tourism is a two-edged sword for island environments. On the one hand, it supports the local economy with income and jobs, reduces pressure on exploitation of other resources, and requires some level of protection of the natural environment that the tourist come to enjoy. On the other hand, its success tends to overrun the very resources tourists come to see such as clean air, clean water, wildlife, and pristine scenery. Much-visited coral reefs are damaged by snorkelers, spear fishermen, and the anchors of boats. Popular forests are damaged by soil erosion, accidental fires, trash, and unregulated foot and vehicular traffic. Invasive plants and animals are brought in on tires, in luggage, and in imported goods. Tourism is also a drain on limited resources, directly through the use of water and space by tourists and indirectly through the impact of a large infrastructure that is needed for support personnel and services. Issues of access to natural areas, such as beaches, appropriate scale and location for resort development, and maintenance of aesthetic "view sheds" all become important. Centers of tourism often develop in cultural and economic isolation from the local culture, re-interpreting it in ways that may lack much basis in reality. Ecotourism, purportedly gentler on the environment, and sometimes initiated to conserve natural areas, still utilizes the same resources, albeit ideally to a lesser degree. Frequently ecotourism expands into previously remote habitats that are still pristine – at least until the tours start visiting them. Whatever form it takes, tourism continues to rise. The nine island groups received 64 million tourists in 2009, 7% of the 900 million tourists that travel every year.

Islands have a particular challenge, then, to balance the often conflicting goals of economic development (often erroneously considered synonymous with the development of tourism) and conservation of the natural resources upon which all nature-based tourism relies. Tourism increased on all the nine island groups between 1950 and 2000, although it was slower to develop in Tonga and Iceland than on other island groups, such as the Canary Islands and Hawai'i. For example, in 1997, Tonga had only one hotel room per square kilometer of area and only an average of 11 visitors per 1000 residents each day, each of whom spent an average of USD 119 per visit. In Hawai'i in 1997, there were four rooms per square kilometer, 134 visitors per 1000 residents each day and an average of USD 1127 was spent per visit. In the Canary Islands in 1997 there were 52 rooms per square kilometer and 164 visitors per 1000 residents each day, each of whom spent on average USD 2740 per visit. In this section, we discuss three island groups (Canary Islands, Jamaica, and Hawai'i) where tourism increased dramatically between 1950 and 2000 and where revenues from tourism constitute a substantially higher percentage of the GDP (at least 15%) than for the other island groups (less than 5%).

Despite some visitation by British tourists in the beginning of the twentieth century, the Canary Islands were not only physically but also

Table 7.2. *Development in the Canary Islands between 1960–2006. From Fernández-Palacios and Whittaker (2008). M = million; K = thousand*

Parameter	1960	2000	2006
Population (M)	0.94	1.78	1.99
Number of tourists (M)	0.07	12.0	12.5
Population density (people per km²)	130	231	266
Cultivated area (K ha)	95	46	46
Number of cars (M)	0.02	1.08	1.3
Income per capita (K USD)	4.3	17.2	25.8
Literacy (%)	36.2	96.4	No data
Female life expectancy	65	82	83

culturally and economically isolated from mainland Europe. Less than 40% of the population was literate and agriculture for both subsistence and export was the basis for the economy. Thousands of people had emigrated to Cuba until the 1930s and then to Venezuela when Spain was economically depressed after World War II. However, by the 1960s, the sun-drenched Canary Islands turned out to be a particularly popular destination for vacationing Europeans (particularly Scandinavians) and tourism quickly replaced agriculture as the main source of income. The four easternmost islands (Lanzarote, Fuerteventura, Gran Canaria, and Tenerife) have been totally transformed by tourism, even to the extent of bringing in sand from the Sahara Desert in order to make the lava-strewn southern coast of Tenerife more tourist-friendly. The three westernmost islands (La Gomera, La Palma, and El Hierro) have retained more of their rural, agricultural base. Table 7.2 shows some of the social and demographic changes wrought by tourism-driven development between 1960 and 2006. A doubling of the population has been due more to immigration (sometimes from returning emigrants from South America, sometimes from African refugees) than local reproduction. A more than 12-fold increase in the number of tourists has meant that a population of nearly 2 million today is supplemented by 300 000 tourists daily for a net population of 2.3 million. Tourism currently accounts for at least 32% of GDP (depending on how calculations are made). Densities on the most populated islands of Gran Canaria and Tenerife reach almost 500 people per square kilometer. Positive results include increased income, literacy, and longevity of Canary Islanders, and decreased grazing pressure on forests. Negative impacts, especially in the arid lowlands where tourists congregate,

Fig. 7.13   Cruise ship and hotels at Ocho Rios, St. Anne, Jamaica.

include increased demands on limited water supplies for hotels, golf courses, and residences; and marine pollution from siltation and sewage. The driest islands, Lanzarote and Fuerteventura, which now receive about 80% of their revenue from tourism, have tackled severe water shortages with desalination plants.

Jamaica attracts 3 million tourists each year to its 26 000 hotel rooms to play tennis, golf, ride horses, and enjoy the beaches and coral reefs (Fig. 7.13). Jamaica advertises its lush natural areas including forests with many species of birds, orchids, and ferns. The tourist industry employs at least 25% of the population and contributes about 20% of GDP. Tourism began in the 1890s when United Fruit Company encouraged cruises to Jamaica in order to use the excess capacity of its ships. However, following World War II, governmental assistance helped spur tourism and the number of hotels tripled between 1945 and 1970. About three-quarters of all tourists are from the USA and, despite a drop in the late 1970s, tourism has continued to flourish.

In Jamaica, most tourists go to coastal resorts that are often owned by foreigners and cater strictly to tourists, not locals. These mega resorts are often in walled off compounds where interactions with the local population are discouraged. This separation can lead to resentment by the locals based on the restrictions placed on them by the hotel chains. For example, certain stores and beaches are considered off limits to all but hotel guests – a status few Jamaicans can afford. When interactions do occur they are typically when locals peddle wares for sale. This peddling can be incessant, driven at times by high local unemployment, but tourists tend to feel harassed. Some locals find employment, either part-time as builders of the resorts or as providers of low wage services such as cleaning and food preparation. Poverty in local villages furthers the contrast with the wealthy hotels but together, both rich and poor contribute to coastal pollution and sedimentation of reefs. Thus, the rapid development of coastal areas harms the very product that the resorts sell – clean beaches and healthy coral reefs. Further, years of tax breaks for resort builders result in little income returning to Jamaica from these resorts. This eventually self-destructive promotion of unfettered development has been discussed by those promoting sustainable tourism in Jamaica and elsewhere. Sustainable tourism promotes reduced consumption of water and energy, lowered waste production and pollution, appropriate technology, and recycling. Sometimes this involves redesigning old structures rather than spending the resources to construct new ones. The University of West Indies Institute for Hotel and Tourism is currently making efforts to measure the carrying capacity of popular areas such as Montego Bay and Ocho Rios. However, tourism as it is now practiced in Jamaica is not sustainable.

In Hawai'i, tourism contributes about 20% of GDP and Hawai'i was a pioneer in developing the "sun and beach" model of island tourism. Famous writers that visited the islands in the nineteenth and early twentieth centuries include Mark Twain, Herman Melville, Robert Louis Stevenson, and Jack London. These and other famous visitors helped raise interest in visiting Hawai'i. By 1929, there were already 22 000 tourists per year but a rapid increase occurred following the first jet airplane flights to Hawai'i in 1950 (bringing 50 000 visitors to the islands that year). When Hawai'i became a state in the USA in 1959, tourism got a further boost and by 1967 over 1 million tourists visited Hawai'i each year. Rapid growth in visitors increased until the late 1980s and since then numbers have been fairly stable between 6.5 and 7.5 million tourists per year. As tourism has evolved, first on O'ahu, then Maui, the Island of Hawai'i, and Kaua'i, it has focused largely in concentrated zones such as Waikīkī on O'ahu, Lahaina and Kihei on Maui, Po'ipu on Kaua'i, and Kona on the Island of Hawai'i (Box 7.2 and Fig. 7.14). Tourism has sometimes replaced failing agricultural industries. On Lāna'i, when the pineapple industry waned in the 1980s, two luxury hotels and championship golf courses were built. Although tourists are now exploring throughout each of the islands as rental cars, bed and breakfast inns, and resorts in smaller towns become prevalent, some stretches of coastline remain relatively undeveloped, at least for tourism. Many tourists now seek alternatives to hotels such as timeshares or condominium vacation rentals, making their environmental impact difficult to distinguish from that of more long-term residents.

Box 7.2. Two sides of Hawaiian tourism

When taking a vacation on Oʻahu, Hawaiʻi, there are at least two distinctly different types of cultural experiences you can have: the mainstream tourist experience and the one off the regular tourist route. Living in Hawaiʻi during the past year, I have tried them both. For the mainstream experiment, you can do what most tourists do: rent a room in a high-rise hotel on Waikīkī Beach in Honolulu (Fig. 7.14); shop at high-priced shops along Kalakaua Avenue one block back from the beach; take an open-air bus tour of Honolulu; lounge in the sun in a rented beach chair; go to dinner and be serenaded by live, sultry Hawaiian music at an open-air restaurant on the waterfront; then catch a show at one of the large hotels or watch a movie on an outdoor screen right on the beach. If you are the active type you might climb nearby Diamond Head Volcano; surf Waikīkī's gentle waves; golf along the seashore; or fish offshore for marlin or tuna. You will never be far from your hotel, your favorite food chain, or your fellow tourists, but you will relax, see great sunsets, and take home a suntan. However, getting off the regular tourist route increases your chances to meet locals and learn about their concerns, connect with nature, and develop your own perceptions of Hawaiʻi without the tourist façade. You can catch a glimpse of Hawaiʻi's human history at the Bishop Museum; learn about both native and introduced plants at Lyon Arboretum; sample locally grown produce such as papayas, mangos, or macadamia nuts at a farmer's market. You can venture out to eat malasadas (a Portuguese donut); catch a local hula group practice in Kapiolani Park; watch a festival in Chinatown; or talk with local fishermen about their catch. This dichotomy of cultural experiences continues if you visit other islands. If you want more classic tourist amenities that repeat Honolulu, then head to the cities of Kona (Island of Hawaiʻi), Kahaina (Maui) or Lihue (Kauaʻi). If you want a less tourist-oriented experience, head to the Hilo side of the Island of Hawaiʻi or to Molokaʻi. LRW

Fig. 7.14   Waikīkī, one of the world's most famous beaches and a center of Hawaiian tourism (see Box 7.2).

Hawai'i, along with other mature destination resorts such as the Balearic Islands of Spain (including Mallorca and Minorca), has begun to face the issues of sustainability of their tourism industries. Hawai'i has pioneered land zoning for urban, rural, or conservation use with sub-zones reflecting environmental sensitivity. This carefully regulated land usage dates back to the time of Hawaiian royalty. Shoreline development is carefully controlled and new programs were instituted in the 1970s for additional coastal protection and in the 1990s to include environmental impact assessments for any potential coastal development. Hawai'i has instituted a hotel room tax that has gone toward the enhancement of tourism quality and sustainability. Short- and long-term planning concerns include the environmental well-being of the islands and the regulations and sometimes restrictions placed on tourism to limit damage to natural resources. These steps bode well for the future of Hawai'i's tourism. Hawai'i also has the opportunity to greatly increase the involvement of native Hawaiians in future decisions about tourism. Such involvement would most likely result in more environmental awareness, particularly if future tourism development retains strong cultural and rural elements, perhaps using Moloka'i rather than Waikīkī as a model (Box 7.2). However, both the current mass tourism model and a potentially more sustainable model of Hawaiian tourism face future challenges from the rising cost of jet fuel, competition from other tourist destinations, a high cost of living, and the steady erosion of natural resources by a growing resident and tourist population.

Tourism is the world's largest industry, with an infrastructure worth almost USD 3 trillion. The environmental impact of such an industry is huge. Sustainable or responsible ecotourism is an effort to reduce the ecological footprint of tourism, forge stronger ties with local economies and communities, and explore alternatives to traditional modes of travel that often sustain unjust governments, multinational corporations, and cultural monotony instead of promoting local entrepreneurs and diversity (Box 7.2). Islands are very popular with tourists and with the finite territories of islands, local governments have an unusual degree of control over access and abuse of natural areas. However, island governments are often poor and heavily reliant on outside income, particularly from tourists, to sustain even a basic standard of living.

The future of tourism will be a race to see if lovely island retreats are spoiled before safeguards are put in place to protect them. A first step in protecting natural areas is to limit access to them. Establishing reserves and parks with variable forms of limited access is one approach. Road closures can restrict tourist access to protect rare wildlife such as the Puerto Rican parrot in the Caribbean National Forest in Puerto Rico; trailhead restrictions can limit hiker access to such popular trails as the Milford Track in New Zealand; and a combination of entry fees, limited opening hours, and limited parking can reduce pressure on popular snorkeling spots such as Hana'uma Bay in Hawai'i. Small islands off larger islands also offer some protection such as the wildlife refuges on Kapiti Island off New Zealand and Culebra Island (and numerous cays surrounding it) off Puerto Rico. The Icelandic government restricts access to Surtsey to only a few scientists each year (see Box 2.2). Ocean-based tourism is important on all of the nine island groups (whale watching provides up to 12% of income in parts of rural Scotland) and provides some inherent limits to access. Finally, scenic flights such as to the fjords and mountains of southern New Zealand result in less impact on the land but have environmental costs that ultimately make them unsustainable (see Chapter 8). None of these methods of limiting access is perfect and overuse can still impact natural resources. Indeed, helicopter tours can severely impact the quality of a wilderness experience for land-based tourists. An important second step is to ensure that funds from tourism remain on island and help fund the environmental costs of tourism, including waste management, education programs for visitors, and support personnel such as guides, rangers, and wardens. Finally, island attractions, which are the basis for all tourism, need long-term conservation and management, particularly when the impacts of natural disturbances on the landscape are factored in (see Chapter 3). Sometimes natural disturbances provide tourist attractions, such as with the highly popular and relatively accessible lava flows on the Island of Hawai'i. More often disturbances deter tourists or damage necessary infrastructure, such as earthquakes in Japan, cyclones in the Caribbean, and drought in the Canary Islands. Long-range environmental planning is needed to cope with both the expected and unexpected impacts of tourism on island resources.

7.5.2 Sport hunting

Introduction of domesticated animals to islands has been a human survival tactic for centuries (see Chapter 5) but introducing big game animals for public sport hunting is a more recent phenomenon. Sport hunting overlaps with hunting for food (as with hunting of feral pigs in Hawai'i and New Zealand), but the expansion of hunting has been due more to increased wealth and leisure than to technological advances. Both domesticated animals and sport hunting have competed with native island animals (where they exist) and altered resources of local island environments. However, sport hunting remains popular, so islanders struggle with the dichotomy of promoting yet controlling populations of game animals. On large islands with low density human populations such as Hawai'i and New Zealand, sport hunting is least restricted. Big game hunting is more restricted on large, populated islands (Japan, the British Isles) and least common on small islands (Tonga, Canary Islands, Puerto Rico, and Jamaica). We discuss three islands where game animals were all introduced (Iceland, Hawai'i, New Zealand), then briefly note two islands (Japan, the British Isles) where humans and game animals have long co-existed.

In Iceland, wild Norwegian reindeer were introduced in the 1770s but by 1817, fearing that their growing numbers were destroying upland pastures normally used by sheep, unlimited hunting was allowed. Then, due to decreasing numbers of reindeer, hunting was restricted in 1882 and banned from 1901 until 1940. Today, a protected population of about 3000 reindeer lives in eastern Iceland but some culling is allowed each year to control their numbers.

Hawai'i has been subjected to many more introductions than Iceland. Various mammals had been introduced by Polynesians (see

Chapter 5) and Europeans (see Chapter 6). Pigs were introduced by Polynesians and further introductions were made by James Cook in the eighteenth century. Pigs are still a popular hunting animal in Hawai'i. Axis deer were the first animals brought to Hawai'i specifically for sport hunting. They were introduced during a 92-year period (1867–1959) by the Hawaiian government. Mouflon sheep (1950s), pronghorn antelope (1959), and mule deer (1961) were also introduced specifically for recreational hunting. All but the pronghorn are still thriving. The Corsican mouflon survives in wild populations on the Island of Hawai'i. A population of mouflon-feral sheep hybrids was introduced to Mauna Kea Volcano on the Island of Hawai'i in 1962 and mouflon to Mauna Loa Volcano in 1968. When growing mouflon populations threatened the highly edible but rare Mauna Kea silversword plants (see Box 4.1 and Fig. 4.2), attempts were made to eradicate the mouflon. However, despite two court orders to remove the mouflon and several decades of effort, the mouflon survived. Silverswords are now only safe within fenced areas.

New Zealand has had perhaps the largest number of species introduced for hunting. The list includes moose, six species of deer (red, sika, rusa, sambar, white-tailed, and fallow), two goat relatives (tahr, chamois), and pigs. While most introductions were in the 1860s to the 1910s (see Chapter 6), populations and attitudes toward these game animals are still in flux. For example, in the 1980s a moratorium was placed on hunting the sambar deer because its survival was presumably threatened, but that moratorium has since been lifted. For 23 years after its introduction, the chamois was protected, but since 1937 there have been no restrictions to hunting it. The Himalayan tahr, introduced between 1904 and 1919, has been actively hunted in order to control its populations but still has an extensive range throughout the mountains of the South Island. New Zealanders seem torn between awareness of the damage these animals are doing to native plant communities and the desire to have healthy herds for recreational hunting. Arguments for continuing hunting include maintenance of the hunting tradition, income from hunters, and control of animal populations. Arguments against hunting include the concerns of conservationists seeking to preserve the high biodiversity on the islands and the drive to gradually remove all introduced species from New Zealand.

The much older human cultures of Japan and the British Isles have developed with resident game animals. In Japan, the native sika deer (shika is Japanese for deer) are still found in the wild, mostly on

Hokkaido, where they were almost hunted to extinction in the late 1800s. Now they are protected, with some hunting allowed to control their populations, and have rebounded to the point of hyperabundance since 1950 (see Chapter 6). Many individuals are tame and are found wandering among temples and in city streets. Sika deer have been introduced to many other islands, including New Zealand and the British Isles. In the British Isles, in addition to sika deer there are five other types of deer (red, roe, fallow, muntjac, and Chinese water), which are hunted for sport. When populations grow too large they are culled. In both Japan and the British Isles, the loss of natural predators (wolves), restrictions on hunting for safety or conservation purposes, and a general reduction in interest in sport hunting have contributed to the growth of deer populations, often to the detriment of the environment.

Hunting also occurs for non-recreational purposes. In Puerto Rico, government-run hunting began in 2008 of the more than 1000 descendants of patas monkeys and rhesus monkeys used in medical experiments in the 1950s. Concerns from local farmers, health officials (some of the monkeys carry herpes and hepatitis), and urban residents in San Juan have prompted culling of individuals not wanted by zoos. However, the likelihood of ridding Puerto Rico of its unwanted monkeys seems remote.

7.6 INVASIVE SPECIES

7.6.1 Dispersal

The dispersal of species around the world is a natural event. Indeed, such processes resulted in the early colonization of the island groups (see Chapter 4). An invasive species is one that successfully establishes at a new site and then spreads into that new habitat. A native species is one that invades an island without transport by humans, while a non-native species (some of which are invasive) requires transport (intentional or not) by humans. Humans have thereby greatly accelerated this process of invasion, particularly on islands that are vulnerable to invasive species (see Chapters 5 and 6). Non-native species that successfully invade new habitats may or may not out-compete the native species. Although the direction of introduction is often from species-rich mainlands to more depauperate islands, there are also frequent examples of islands serving as sources for species that become invasive on mainlands or on other islands as when New Zealand mud snails invaded the British Isles and Japan. The results of these invasions are a new set of species mixtures and novel ecosystems.

The acceleration of the biological invasion of islands as a result of human activities was particularly pronounced after 1950, when agriculture, industry, and tourism intensified and the amount of air and sea traffic to and from islands increased. Humans have often intentionally introduced crops such as pasture grasses and domestic animals such as cattle and goats. The brushtail possums (see Fig. 1.5) were introduced to New Zealand from Australia for a fur trade and caused serious ecological damage to forests trees and birds (see Chapter 6). Possums have recently been introduced from New Zealand to Japan as pets. So far, they have not, to our knowledge, become invasive in Japan. Hunters often introduce game animals, which then escape from designated areas and spread (see Section 7.5.2). Unintentional introductions have come from ships as stowaways (rats and mice), in ballast water, or attached to ship hulls. Since 1950, animals have invaded islands that had not previously been reached, as when a fishing boat ran aground on Taukihepa off the south coast of Stewart Island in New Zealand in 1964 and ship rats climbed ashore. The rats multiplied and subsequently caused the extinction of the bush wren and greater short-tailed bat. The agricultural, silvicultural, and horticultural industries on islands are also very common vectors for invasive plants, insects, snails, slugs, and soil organisms, especially via potted soil, plant cuttings, and wood products. For example, the coqui frog, which was introduced to Hawai'i from Puerto Rico in the 1980s probably came with plants intended for the horticultural trade. In just a few decades, the coqui population densities have matched or exceeded those in Puerto Rico (up to 90 000 frogs per hectare). Intriguingly, in Hawai'i the coqui's diet consists mainly of non-native ants and non-native amphipods in the forest leaf litter. Escaped pets such as birds, snakes, turtles, or iguanas in Hawai'i and Puerto Rico are additional unintentional sources of invasive species.

7.6.2 Novel biological communities

One inevitable consequence of the global mixing of flora and fauna has been the development of novel biological communities. These are biotic communities that exist only because humans have introduced invasive species. Novel communities can be composed entirely of new species or of mixtures of nonnative and native species. These new communities may preserve many of the ecosystem functions of the original communities (such as nutrient and hydrologic cycles or erosion control). Alternatively, the invasive species may fundamentally alter one or more ecosystem processes. Fire tree provides an example of how an invasive species can alter nutrient dynamics in low-nutrient, Hawaiian uplands (see Box 6.4). Lowland parts of most of the tropical island groups (Hawai'i, Tonga, Puerto Rico, and Jamaica) are now dominated by non-native species. Pasture grasses and other non-native invasives dominate most of the pastoral landscapes of temperate New Zealand, British Isles, and Japan. Even the arid climates of many lowlands in the Canary Islands and the cool climate of Iceland have not prevented the invasion of non-native species, although the proportion of invasive species to native species is somewhat less than on the other island groups. Since the 1950s, the formation of novel biological communities has accelerated with the increase in habitat loss and alteration to support expanded industrial growth. Introductions of new plants for horticulture and animals sold as pets provide additional potential sources of invasive species. Novel biological communities are now integral parts of the landscape and it behooves us to examine them as carefully as the more pristine and increasingly rare native communities.

7.7 RESPONSES

7.7.1 Conservation

The rapid expansion of the human population has caused a great deal of soul-searching among those concerned about overexploitation of resources and potentially irreversible damage to native communities. One response has been the groundswell of support for and involvement in conservation activities. Despite the typical focus on economic growth and job creation, many island residents realized that unbridled growth leads to resource depletion and, in the worst case, to environmental and potentially societal collapse. Islanders have been at the forefront of the conservation movement in several instances, perhaps because they are more aware than mainlanders of the vulnerability of their limited landmass to perpetual growth. Indigenous populations have contributed substantially to island conservation movements, often based on a combination of historical, spiritual, and practical motives. Native Hawaiians, for example, speak of 'aina, or the love of the land. In New Zealand, Māori concern about protecting sacred mountains led to the establishment of Tongariro National Park in 1887

(Table 7.1). The land was gifted to all New Zealanders by Te Heuheu Tukino IV of the Ngāti Tuwharetoa tribe and the park was modeled after Yellowstone National Park founded in 1872 in the USA. In the nineteenth century, deforestation in the Caribbean to grow sugar and other crops led to the formation of local forest services and laws to protect forests covering mountainous watersheds. Industrialization of the British Isles triggered the formation of the National Trust in 1895 to conserve coastlines, countryside, and historical buildings. In the twentieth century, conservation efforts continued. New Zealanders formed the Royal Forest and Bird Protection Society in 1923, campaigned to protect kauri forests from being logged in the 1940s, and coalesced into a national campaign to resist a large hydropower project involving Lake Manapouri in the 1960s (see Section 7.3.3). Subsequent New Zealand actions have included the establishment of marine reserves and a doubling of national park lands. The Japan National Trust, established in 1968, was modeled after the British National Trust to conserve natural and cultural assets. In the Canary Islands, a large national park was established in 1981 to protect upland laurel forests on Tenerife. In the 1990s, Tonga and Jamaica established their first national parks (Table 7.1). Despite dismal records of past conservation, many islands are now recognizing the critical role of natural areas for economic as well as ecological well-being. Efforts are widespread to save what is left. Modern Hawaiian ʻaina is now expressed in many ways, including active participation in conservation organizations, government review boards, and activism that promotes subsistence farming (Fig. 7.15a) and protests further development (Fig. 7.15b).

Establishing national parks and other protected areas is an important step toward conserving natural resources but it is not a panacea. A general problem of designating protected areas is that the designation can result in a partitioned landscape where nature conservation is considered suitable only in a part of the landscape and is scarcely addressed elsewhere (Box 7.3). Many plants and animals do not respect these human-imposed boundaries! In Jamaica, the rate of deforestation within the boundaries of its first national park continued at the same rate over the decade after the park's formation as over the decade before. Unless the causes of deforestation, such as rural poverty and soil conservation on agricultural land outside the park are addressed, this is likely to continue. In New Zealand, a wealthier country than Jamaica, mineral prospecting and mining has recently been proposed within national parks as a means of addressing national debt. Thus, despite widespread popular support for

Fig. 7.15   Local perspectives from rural Molokaʻi. (A) Advice on how to
live sustainably; and (B) resistance to commercial development of wind
farms on Molokaʻi that would supply electricity to urban Honolulu via
undersea cables.

conserving natural resources and clear documentation of declines in New Zealand's unique biodiversity, some New Zealanders see continued extraction of natural resources as a model for its economy for the foreseeable future.

Box 7.3. Ten thousand cranes

Red-crowned cranes in the winter snows of Hokkaido are recurrent image in Japanese art, but these are not the only cranes that spend the winters in Japan. Hooded crane and white-naped cranes migrate from their summer breeding grounds in Mongolia and eastern Russia, down the Korean Peninsula to arrive at the marshes at Arasaki in southern Kyushu for the winter. Arasaki is now the main wintering ground for both of these species of cranes. On a cool January evening, I traveled by train from Kagoshima and arrived in Arasaki after dark, staying overnight at a minshuku (bed and breakfast inn). The next morning dawned clear, with the muted sound of trumpeting cranes audible outside. I saw quite a spectacle outside: more than 10 000 cranes concentrated in a few marshy fields reclaimed from the sea. During the morning, the birds became more active, some flew and circled, but most remained standing and their calling grew in volume. Their habitat is now so reduced that the cranes, which once would have fanned out across the broad, flat plains nearby, are now in the few fields that are suitable, the rest now changed to intensive agriculture. The little remaining habitat is protected as a national monument by the Japanese government, which leases the land from the farmers who own it. The farmers are involved in conservation of the cranes, and I watched the farmers as they drove around the flock of cranes in small tractors, distributing grain for the birds. Despite the restricted habitat and their oddly cramped winter lifestyle, the number of cranes at Arasaki has grown from only 300 after World War II to a current population of over 10 000. PJB

The unbridled growth of human populations and resource extraction has clearly had consequences in the past as we have seen for islands such as Iceland (see Chapters 5 and 6). Modern transportation and economies have enabled wealthy islands to have national parks, and cleaner air and water than in their industrial past (especially in the British Isles), while extracting resources from distant places, including other islands. Less wealthy island countries have been able to develop ways of life maintained by wealth generated elsewhere. For example, in the case of Tonga, the historian Ian Campbell considered that its current population is about three times the maximum number of people on the island before European contact. He suggests that the population's standard of living cannot be sustained by the productivity of Tonga's soils and waters. Instead, that standard is maintained by remittances from migrants and foreign aid; the cessation of either of these would be catastrophic for Tonga's current standard of living. A general step that is needed across all islands is for their residents to be better apprised of the state of their island's natural resources and for there to be general discussion about their island's capacity to sustain the current way of life.

7.7.2 Restoration

Another response to the immense alteration of natural ecosystems by humans has been a growing desire to restore ecosystems. Restoration in the broad sense is human intervention to bring back some degree of original species composition and function to altered ecosystems. An exact re-creation of a habitat that has been disturbed is unlikely. Extinctions caused by humans and, on many islands, wholesale transformation to dominance by introduced plants and animals, are significant impediments to restoration of natural ecosystems on islands. The leeward lowland regions of Kaua'i in Hawai'i are an extreme case in point: here fossil evidence has shown that a forest composed of native trees, shrubs, and birds once existed. Now, nearly all of these species are extinct (see Section 5.7), and in their place is a landscape dominated entirely by introduced species. New disturbances, such as fire regimes begun by humans and perpetuated by introduced, combustible grasses (see Section 6.6.2), can push an ecosystem across a nearly irreversible threshold, and this poses a significant challenge for ecosystem restoration. Even when extinction and invasive species are a less severe problem, the time since deforestation can determine the chance of success. The longer an area has been deforested, the greater the difficulties in restoring the soils and soil organisms. On many of the islands, deforestation of habitats took place centuries ago. Approaches to restoration depend on the damage that has occurred, with likelihood of success being inversely related to the degree of damage.

Goals for restoration vary across and within islands. For some restoration projects, goals may be as simple as restoring forest cover to deforested regions. Achieving this goal may require as little management as the decision to abandon agriculture. For example, abandonment of agriculture in Puerto Rico resulted in invasion of former pasture by a mix of native and introduced trees and the total forested area of the island increased from 9% forest cover in 1950 to 37% in 1990. However, if the goal were to restore forests to maximize the accumulation of carbon (for example, to offset carbon emissions) then this approach may not be ideal. Abandoned fields in Puerto Rico that were invaded by introduced trees, which now form a forest, store less carbon in soils and trees than forests of the same age made up of native trees. Therefore, managing young Puerto Rican forests on former fields for carbon might require intervention to favor native rather than introduced tree species. In New Zealand, abandoned fields are often invaded by introduced shrubs and in many places these former fields are now seen as a step toward restoring native forests because the introduced shrubs are short lived. In some of these areas, introduced grazing mammals are either fenced out (Fig. 7.16) or killed because they might prevent certain native tree species from germinating and growing.

Fig. 7.16   Revegetation on the severely grazed Banks Peninsula, South Island, New Zealand is occurring in small, fenced patches while grazing by sheep and cattle continues over most of the terrain.

Where regions of islands are severely deforested, there is often little original native forest to provide a seed source for restoration. There is also little information about what species comprised former forests. Fortunately there is a growing body of research on fossils, such as pollen, seeds, wood, bones of animals, shells of snails, and even fragments of beetles, than can reveal the composition of forests that no longer exist. These can be used to set goals for restoration. For example, in an upland region in the south of Scotland (Carrifran) the goal for replanting forests is to achieve the same composition the area had 6000 years ago, just before it was deforested. The composition has been determined from analyses of pollen.

On many islands, the goal of restoration is to control the introduced species that threaten the survival of native species. This approach can begin with the removal of introduced plants, especially species that cause new disturbances (such as those that are highly combustible), or which smother regeneration of native plants. It can also involve the control or local eradication of introduced animals (see Section 6.6.1, Box 7.4 and Fig. 7.17). In New Zealand, an approach to restoring the native animal communities in forests within the last decade has been to construct expensive, fine-meshed fences to surround forest fragments: these fences can prevent rats, stoats, cats, and other introduced predators from entering the fenced area. The predators are then eradicated inside the fenced areas by trapping and poisoning. Native birds, lizards, and frogs that are highly vulnerable to these predators are then released into the fenced areas. Some of the fenced forest fragments are small (8 ha Riccarton Bush in the center of the city of Christchurch) while others are large (3400 ha Maungatautari, near Hamilton, surrounded by 47 km of fence). These fences require constant maintenance because of the pressure of reinvasion. Herbivores such as feral goats and predatory mammals such as rats have been eradicated from islands off the coasts of New Zealand, Hawai'i, and the British Isles since the late 1980s. The technology needed to achieve this has developed rapidly and rats have now been eradicated from large islands, including the 11 200 ha Campbell Island, 1000 km south of New Zealand. Soon afterward, a previously unknown species of bird, a snipe, re-colonized the island from a tiny islet offshore and has begun to breed on Campbell Island. Elsewhere, active attempts to restore breeding seabirds to islands from which they were extirpated are underway because of the important role seabirds have in adding nutrients to island soils (see Section 4.4.4). As technology and knowledge are shared, restoring islands is becoming increasingly widespread. Eradication of non-native animals as a means of restoring islands is far from universally accepted. There were public campaigns against the eradication of one of the last populations of ship rats in the British Isles, intended to boost populations of seabirds on the island of Lundy. Debate about eradicating introduced European hedgehogs is on-going in the Outer Hebrides, British Isles, where they eat eggs of wading birds. Goals for restoring populations of seabirds can be divergent among different groups of people. For example, in New Zealand a key goal of restoring small offshore islands for Māori communities is to provide a source of natural resources for people to use, such as populations of seabirds from which chicks can be harvested sustainably for food. For many European New Zealanders, nature preservation is the main goal; in this case to maximize numbers of seabirds.

Box 7.4. Snail jails

Invasive, non-native species are often fenced out of areas that contain threatened native species. Sometimes, however, rare native species are fenced in. During a field trip to examine rat impacts on Oʻahu (Hawaiʻi), Peter and I were introduced to several "snail jails" that do just that – they protect the few remaining native snails in a forest where a non-native predatory snail, called the rosy wolf snail, has invaded. The fence was an elaborate affair (Fig. 7.17), with metal sheeting for sides, then an overhanging top that included electrical filaments, salt barriers, and even some token barb wire on the top (for the really big snails?!). Unfortunately, the populations of native snails within these 6 by 10 m enclosures were not doing well, perhaps because of a loss of genetic diversity among the small populations, or because rats are not completely stopped by the fence, or because of the death of the native vegetation within the fence that provides food for the snail. LRW

The challenges of restoring heavily altered forests are formidable. Restoration of Kahoʻolawe, Hawaiʻi is underway after decades of use of this island for bombing practice by the US military. About a third of the island still has not been cleared of unexploded ordnance, limiting human access for restoration purposes. Other heavily bombed areas in need of clean up include Okinawa, Japan, where an estimated 2500 tonnes of unexploded bombs remain. Another challenge is to

Fig. 7.17   Snail enclosure to keep native Hawaiian snails safe (inside the structure) from non-native snails outside the structure. The metal siding is topped by salt and electric barriers (see Box 7.4).

restore formerly fertile Icelandic pastures that are now mostly gravel beds that are unable to sustain plant growth (see Box 7.5, Fig. 7.18 and Fig. 5.10). Deforestation of islands often is difficult to reverse because, without forests, wind speeds increase, humidity decreases, and conditions for plant growth become difficult. However, successful replanting of forest species, as done for some mid-elevation pine forests in the Canary Islands, can lead to immense benefits, including stabilization of soils, preservation of water tables, and protection of habitat for understory plants and animals. Therefore, among the wealthier islands we consider, restoration to reduce the rate of degradation of soils and native plants and animals is a growing activity, often undertaken by citizen groups as well as their governments. For poorer countries, such as Jamaica, it is a challenge to find the resources to prevent further deforestation and restore degraded habitats.

Box 7.5. Mixed blessings in efforts to restore Iceland's soil

The Soil Conservation Service in Iceland had a good idea: import an Alaskan lupine to help restore fertility to Iceland's severely eroded soils on farms. The lupine grew well and added nitrogen and organic matter to the soil. Many farmers eagerly planted the seeds and watched a bumper crop of lupine cover their barren fields. However, about ten years later they still had dense fields

Box 7.5. (cont.)
of lupine. The lupine is toxic to sheep and cattle. Lupine seeds also remain viable for decades in the soil, and lupine inhibits slower-growing native plants and native forage plants. Farmers who once were eager to plant lupine now ask for advice on how to get rid of it. This dilemma reminds restorationists that what one introduces as the immediate solution may become a longer-term problem (see Fig. 7.18 and Fig. 5.10). – LRW

A

B

Fig. 7.18   Revegetation in Iceland's barren wastes. (A) lupine plantings in rows in formerly vegetated, upland pastures; and (B) close-up of a lupine field (see Box 7.5).

7.7.3 Environmental limits imposed on humans by island ecosystems

Throughout this chapter we have focused on human effects on islands but underlying each of these are, of course, the physical and ecological restraints that islanders face. The finiteness of islands limits the resource base on which islands can draw, thus naturally limiting human population growth. An obvious constraint is space, both for living and for agriculture. Of the island groups that we consider, all but the British Isles and Tonga are mountainous (Fig 7.19). Such rugged terrain is unsuitable for agriculture, which then is limited to relatively small, coastal floodplains. The result is that many of the islands are not self-sufficient and must import food. The physical isolation of islands limits human exploitation of resources. Large inputs are needed from mainlands, such as electricity, labor, or specialized equipment. Mining companies, for example, need to consider the expense and logistical difficulties of initiating work on an island and hauling minerals to other parts of the world. Isolation also requires special arrangements for perishable crops such as lettuce, meat, or dairy products, although the advent of refrigerated units helped, particularly for the New Zealand lamb and dairy industries. Modern transportation facilities now allow the expansive growth of other global industries in cities such as Honolulu,

Fig. 7.19   Tunnel for road at Arthur's Pass, South Island, New Zealand.

Hawai'i (tourism) or Nagoya, Japan (shipping). Each of these economic activities has numerous environmental limits. For example,
water for agriculture, industry, and tourism is limited and heavy
use can deplete local aquifers; ports have limited room for expansion and are constrained by the topography of the coast; and land
for waste disposal is limited and waste can contaminate aquifers.
At a personal level, humans on islands are limited in travel and job
opportunities as well as in the development of recreational areas.
Only with the advent of jet travel has there been a wide variety
of foodstuffs available in island stores. Native-born islanders and
those choosing to adopt an island life style generally sacrifice the
mobility and range of choices on mainlands in exchange for other
benefits of island life.

7.8 STATE OF THE ENVIRONMENT IN 2000

The state of the environment was generally worse in 2000 than in
1950, despite a few bright spots such as increases in the size and
number of national parks and a more robust environmental movement. The damage from World War II had largely been repaired but
the expansion of military bases, particularly those of the USA, on
Japan, Hawai'i, Iceland, and Puerto Rico occupied substantial territory. Global capitalism was the central economic model in 2000 and it
entails the unbridled use of resources. Fortunately, natural resources
were no longer seen by a majority of island residents as expendable
commodities, particularly as prices of such resources as lumber, oil,
fish, land, and water had all increased and as the supplies of some,
such as North Atlantic cod, had been exhausted. This shift in attitude
was due, in part, to a growing realization of the limits of growth.
Annual State of the World reports reflected the preoccupation with
emerging environmental problems and resource limits, as did an
increasing number of international agreements on environmental
themes. However, most of the resources that were depleted on the
island groups were merely imported from other parts of the world.
In fact, by 2000, most islands imported most of their food and fuel.
Agriculture had reached a new level of efficiency by 2000 with max-
imal use of fertilizers, drip watering systems, and highly mechanized
harvest methods on the richer island nations such as Japan, New
Zealand, Hawai'i, and the British Isles, but local farm ownership and
more traditional methods still dominated in poorer nations such as
Tonga and Jamaica. The more mechanized farmlands also generally

had poorer records of maintaining native ecosystems. While agribusiness expanded its reach, the total land area devoted to agriculture generally decreased, particularly in Puerto Rico, where urbanization resulted in wholesale land abandonment. A number of crop shifts occurred in the latter part of the twentieth century because of market forces and local resource depletion.

Island tourism grew exponentially between 1950 and 2000. Massive development along coastlines and the era of big resorts transformed many island communities. Domestic and international tourism was nearly global, with middle classes mixing with the wealthier citizens and an island holiday becoming accessible to a much larger fraction of mainland populations. The degree of urbanization increased along with population increases, with more than half of the world's population living in cities by 2000. Islanders migrated to cities that grew apace with mainland cities. Many mainland cities were populated in part by islanders. Yet the urban footprint continued to threaten island resource bases and the doubling or tripling of island populations since 1950 offset most efforts at conserving natural areas.

Invasive species were much more widespread by 2000 than in 1950, largely because of five decades of increasing levels of international travel and commerce. Many novel ecosystems challenged ecologists with new characteristics, including negative ones, such as a tendency to burn or erode, and positive ones, such as the stabilization of slopes or the creation of shaded conditions that can help the re-establishment of some native plant species. The best outcome of the intensification of human impacts was the development of a strong environmental movement that was global in scope, with both local and national or international leadership, and flexible enough in its approach to address many problems such as deforestation, water pollution, and deep-sea mining. Was the global environment in a better place in 2000 than in 1950? Probably not, as a doubling of the human population in a world of finite resources left increasingly little room for maneuvering around environmental problems. Island problems reflected mainland problems but were, if possible, even more severe because of the finite nature of space and resources on islands. Yet, the outlook for humanity in general and island resources and cultures in particular was not all gloom and doom. Human societies did not appear to be in danger of imminent collapse and the prospect of nuclear war had lessened. We explore the current environmental issues on islands and look toward the future in the final chapter.

SELECTED READING

Carson, R. (1962). *Silent Spring*. Boston, MA: Houghton Mifflin.
Denslow, J.S. (2003). Weeds in paradise: thoughts on the invasibility of tropical islands. *Annals of the Missouri Botanical Garden* **90**: 119–127.
Douglas, I., R. Hodgson and N. Lawson (2002). Industry, environment and health through 200 years in Manchester. *Ecological Economics* **41**: 235–255.
Earthtrends (2009). http://earthtrends.wri.org (accessed 21 October 2009).
Ehrlich, P.R. (1968). *The Population Bomb*. New York: Random House.
Eyre, L.A. (1987). Jamaica: test case for tropical deforestation? *Ambio* **16**: 338–343.
Fernández-Palacios, J.M. and R.J. Whittaker (2008). The Canaries: an important biogeographical meeting place. *Journal of Biogeography* **35**: 379–385.
Grove, R.H. (1995). *Green Imperialism: Colonial Expansion, Tropical Island Edens and the Origins of Environmentalism, 1600–1860*. Cambridge: Cambridge University Press.
Hall, C.M. and S. Boyd (eds.) (2005). *Nature-based Tourism in Peripheral Areas: Development or Disaster*. Clevendon: Channel View.
Harada, K. and G.P. Glasby (2000). Human impact on the environment in Japan and New Zealand: a comparison. *The Science of the Total Environment* **263**: 79–90.
Jackson, J.B.C., M.X. Kirby, W.H. Berger, *et al.* (2001). Historical overfishing and the recent collapse of coastal ecosystems. *Science* **293**: 629–638.
León, C. and M. González (1995). Managing the environment in tourism regions: the case of the Canary Islands. *European Environment* **5**: 171–177.
Leopold, A. (1949). *A Sand County Almanac*. New York: Random House.
Lugo, A.E., L.M. Castro, A. Vale, *et al.* (2001). *Puerto Rican Karst – A Vital Resource*. United States Forest Service General Technical Report WO-65. Washington, DC.
McCaskill, L.W. (1973). *Hold This Land*. Wellington: Reed.
Sheldon, P.J. (2005). The challenges to sustainability in island tourism. Occasional Paper 2005-01, School of Travel Industry Management, University of Hawai'i at Mānoa, Honolulu, Hawai'i.
Sheldon, P.J., J.M. Knox and K. Lowry (2005). Sustainability in a mature mass tourism destination: the case of Hawai'i. *Tourism Review International* **9**: 47–59.
Tsutsui, W. (2003). Landscapes in the dark: toward an environmental history of wartime Japan. *Environmental History* **8**: 294–311.

8

The future of island ecosystems:
remoteness lost

8.1 INTRODUCTION

As biologists we are trained to observe natural patterns in the environment and examine how these patterns change over time. Yet, whenever we attempt to make predictions about future changes, inevitably we are humbled by the variability in ecosystem responses to disturbance. On the one hand, island ecosystems are under great strain from increasing human populations, global climate change, environmental destruction, and exploitation of natural resources. On the other hand, some island ecosystems rebound remarkably quickly when left undisturbed, especially if critical thresholds have not been reached. Predictions are further complicated because economic forces can both exacerbate and reduce impacts of disturbance, and human intervention can sometimes successfully address challenges such as erosion control, invasive species, and habitat restoration. Therefore, precise predictions about the future of any given ecosystem are unlikely. Despite variation across the nine island groups in size, geology, climate, disturbance history, and human impacts, some common patterns of ecosystem responses to disturbance emerge that suggest a common future. In this chapter, we present our views of some of the challenges that face island ecosystems and some of the constructive approaches being used to address these challenges. We then consider the broader implications of lessons we have learned from the nine island groups.

8.2 POPULATION PRESSURE

Human survival depends on adequate supplies of resources but, especially on islands with their finite space, resources are not usually sustainably harvested. Modern island societies have overcome such

natural limits by importing most of their resources, leading to very
dense human populations, as in Japan and Puerto Rico. As modern
societies are challenged to keep citizens healthy and employed, human
numbers keep growing, thereby increasing the challenge. Growth
without limits is impossible, so sooner or later human societies must
stop growing in response to a lack of resources. When those limits
are approached in an uncontrolled way there can be undesirable side
effects including famine, war, and disease. Alternatively, societies can
address the resource challenges before they become too severe, allowing for a more desirable transition to a zero growth economy. The bubble of prosperity during the last half century of optimism depended on
the belief that growth was inevitable, indeed necessary, for a healthy
economy. Modern prosperity has relied on the fact that resources
were easily exploitable and apparently unlimited. Unfortunately,
from a biological perspective, too much growth in numbers of any
species in a community is neither inevitable nor desirable – at least
when resources are exhausted. A re-evaluation of economic theory is
occurring along with renewed emphasis on environmental concerns,
but this recent perspective is still a minority view compared to economic orthodoxy and the belief that there will always be technological
solutions. From an environmental health perspective, we have high
population pressure, which is a warning that our environments are
susceptible to further problems.

Over one-half the world's human population lives in cities, and
island populations are no exception. Yet modern cities such as Nagoya,
Japan, or London, British Isles were built on flat, coastal floodplains
where construction was easy, fertile soil for crops to feed growing
populations was nearby, and ocean transport was accessible. In some
cases, in-filling of coastal wetlands allowed urban expansion, as in
Honolulu, Hawai'i. As cities expand, however, critical agricultural
land is lost and urban residences become increasingly susceptible to
flooding and silt deposition – the processes that initially formed the
floodplain. Rivers can be constrained in the short term (Fig. 8.1) but
floods, like other natural disturbances, ultimately dictate where people live.

Island populations are not equal in how they use resources.
Citizens of wealthy countries such as New Zealand, the UK, and
Japan make a larger demand on the world's environment (have lar-
ger ecological footprints) than citizens of poorer countries such as
Tonga and Jamaica. Most of us are vulnerable to the desire for the
trappings of wealth, especially if it can make our lives healthier and

Fig. 8.1   A river in the city of Nagoya, Japan constrained by concrete banks.

more comfortable, and expand our opportunities to work, study, or travel. Therefore, the citizens of the world generally desire what the richer citizens have. If estimates are correct that we are now using most of the world's resources, the increase in resource use as the poorer two-thirds of the world become richer could demand the equivalent resource pool of two more planets. Island nations are so dependent on imported resources that they will be among the earliest affected by resource shortages. The Earth can sustainably support somewhere between 2.5 and 12 billion people (depending on how one defines "sustainable"). Now that we are approaching 7 billion people, and projected to reach 9 billion by the middle of the century, how will we know when the Earth has reached its capacity for supporting humans? If current human populations have exceeded that capacity, the implications are dire. If there is room for a few billion more humans, how will we adjust to the increase, given our current environmental challenges? Very few societies are directly addressing issues of overpopulation, such as through economic incentives for having smaller families, but successful indirect methods include increasing education and reproductive choices available to women.

Island societies, rich or poor, confront growing populations and resource limitations in different ways. The UK and Japan are economic powerhouses with high population densities that maintain their status and numbers by drawing a large per-capita share of global resources. Yet there are clear indications that sustaining this standard of living is becoming untenable. The UK has a very large national debt, and the rate of growth of Japan's economy (the "bubble economy") began to falter and stall from the 1990s onward. New Zealand and Iceland maintain a high standard of living, but have done so in part by pillaging their once ample natural resources and in part by strictly controlling immigration. The Canary Islands, Puerto Rico, and Hawai'i have maintained burgeoning populations despite declining natural resources because of strong economic support from mainland nations (Spain for the Canary Islands and the USA for the other two). Independent Tonga and Jamaica have high population densities and diminishing resources and are heavily dependent on remittances from emigrants. Many of the islands' resources have been depleted because of successive export crops yet human populations have only grown, largely because of increased global dependency. Human populations on islands and their interactions with island environments are therefore excellent laboratories to study population pressures because they have a clearly finite pool of local resources and their interdependence on global resources is relatively easy to assess.

8.3 CLIMATE CHANGE

Aside from population control, global climate change is likely the biggest environmental challenge humans face. Tackling it will require fundamental changes in resource use, but life styles are the last thing most of us want to change. Climate change is not to blame for all environmental problems, but it interacts with other anthropogenic disturbances to alter natural environments. What aspect of climate change will be the trigger that promotes a world-wide sense of urgency? Severe drought, flooding of our homes, more severe storms, loss of coral reefs, expansion of tropical diseases, or a tenfold increase in the cost of a plane ticket to our favorite island paradise? No one wants to see island communities lose their vital connections with mainland resources, but what if one day no planes arrive with tourists and no planes leave with export products? Island residents are particularly vulnerable to the impacts of climate change because of the geography, topography, and physical isolation of islands and their reliance on coastal, ocean, and mainland resources. In this section, we discuss a few of the ecological implications of the effects of climate change on temperatures, ocean levels, ocean acidity, coral reefs, and tourism. We also note some positive steps being taken to address these concerns.

8.3.1 Temperature

Steadily increasing emissions of carbon dioxide have been documented since the 1950s at an observatory on Mauna Loa Volcano on the Island of Hawai'i. This and other gases (notably nitrous oxides from automobile exhaust, and methane from cattle and melting permafrost) have reinforced the normal greenhouse effect that keeps the Earth at a livable temperature; therefore, the Earth is becoming warmer. Higher air temperatures may be welcome only in Iceland, New Zealand, and the British Isles but will have multiple, complex, and interacting effects on each of the nine island groups. Increased air temperatures will likely make dry climates (Canary Islands, the leeward sides of Hawai'i and New Zealand) drier, potentially affecting water tables, agriculture, and tourism. The lower elevational limit of forests may rise and lower elevation plant species may be replaced by more drought-tolerant ones. High elevational forests may lose their dependable rainfall and become susceptible to fire. Tropical diseases such as dengue fever and malaria may spread to or become more prevalent on subtropical islands such as Tonga, Puerto Rico, Jamaica, and Hawai'i. Increased ocean temperatures will affect sea levels, coral reef health, algal communities, and invertebrate and vertebrate communities in complex ways. Kelp beds may shift distributions and red algae are thought to be more susceptible to temperature increases than are green algae. As algae shift, fish and other animals dependent on them will be affected. Temperature changes are inextricably coupled with changes in ocean currents, precipitation, and glacial melting. Efforts to reduce greenhouse gas emissions on islands include promotion of electric cars (as in Honolulu) and development of alternative energy sources including wind, sun, geothermal, ocean waves, and tides. Scotland, for example, is planning a massive wind farm off its northern coastline that will eventually supply it with all of its electrical needs. The Canary Islands get nearly one-fourth of their energy needs from solar power and use solar energy in the desalination of sea water. Geothermal energy is used to heat baths in Japan, and greenhouses (Fig. 8.2) and homes in Iceland. Scotland and Ireland are pioneering wave power, and tidal projects are planned along the northwestern shore of New Zealand (Kaipara Harbor) and the southwestern coast of England (Severn Estuary).

Fig. 8.2   Use of geothermal hot water to heat greenhouses in Iceland.

8.3.2 Sea level and acidity

Perhaps the most publicized aspect of global climate change is sea-level rise. Predictions vary from a minimum rise of 20 cm to possible increases of several meters during this century. This rise is projected due to expansion by warmer oceans as well as accelerated melting of glaciers and continental ice sheets. Much of the uncertainty about the amount of rise is because we do not know how fast the ice cap on Greenland will melt. The other major uncertainty lies in not knowing how much humans will control greenhouse gas emissions. Citizens of the Maldives Islands in the Indian Ocean and Tuvalu in the Pacific Ocean, where the land never reaches more than 2 m above sea level, are already planning what to do when their nations are inundated. These countries are urging other nations to begin tough restrictions on greenhouse gases because the threat is very real to them. Most residents of islands live near the coast and sea-level rises predicted by the end of the century would inundate low-lying portions of many large cities such as Tokyo and London. Large tracts of agricultural land and fresh-water wetlands would also be flooded in places such as Jamaica. Salinization of shallow aquifers in porous volcanic and carbonate rocks on many of the islands is also a possibility. With the loss of fresh water, island communities would be severely impacted. Logical reactions to sea-level rise include reducing the source of the problem (greenhouse gas emissions) and preparing for the consequences. The town of Hilo in Hawai'i has not allowed construction along its coastal waterfront primarily because it is vulnerable to tsunamis. A planned housing development on coastal dunes in northern New Zealand was abandoned recently when insurance premiums were too high, given likely inundation following sea-level rises. Elsewhere, coastal zones vulnerable to sea-level rise have been designated as parks or public beaches. However, in many populated areas, little effort is being made to restrict development. Coastal property is so desirable for locals and tourists that construction continues unabated, especially in popular tourist destinations such as the Canary Islands, Puerto Rico, Jamaica, and Hawai'i.

Increased carbon dioxide in the atmosphere means more dissolved carbonic acid in the oceans. This increased acidity is almost certain to hamper the ability of marine organisms, such as the development of coral reefs by corals. Higher acidity in the ocean could also eventually lead to increased alkalinity as calcium-based material dissolves into the ocean. That alkalinity might eventually lead to more absorption of carbon dioxide by the oceans but not before extensive damage to marine life has occurred.

8.3.3 Coral reefs

The benefits of coral reefs in tropical seas are indisputable, but the survival of reefs is in doubt. Coral reefs provide spawning and feeding grounds for many fish, including over half of those regularly consumed by humans. Reefs also maintain sandy beaches by absorbing energy from waves, and those beaches are vital to the tourism industry. The algae living in coral even emit a gas that, when it reaches the air, helps clouds form over coral reefs, maintaining a cooler ocean temperature. The loss of coral reefs is not a hypothetical problem. About 50% of reefs in the Caribbean are already gone (about 19% globally). An additional 15% could be gone within 20 years. Direct threats to coral reefs include ocean level rise; higher ocean temperatures and acidity; pollution; sediments from eroding agricultural lands; trawling of ocean floors or dynamiting by fishermen; recreational damage from boaters, divers, and snorkelers; and trade in coral jewelry. The fate of corals depends on curbing as many of these threats as possible. Reefs can begin to recover rapidly when declared off-limits to fishing. For example, O'ahu (Hawai'i) has instituted some annually rotating limits on shore fishing; yet less than 1% of Hawai'i's coastal waters are completely protected. Estimates are that to preserve the 75% of Hawai'i's fish populations that are depleted, 20 to 30% of coastal waters need protection. Fortunately, marine sanctuaries that restrict commercial fishing are increasing in number and size. In 2006, the USA established the 363 000 km^2 Papahānaumokuakea Marine National Monument in far northwestern Hawai'i (a surface area larger than all US national parks combined) and the British government announced in 2010 that it is planning a 544 000 km^2 reserve around the Chagos Islands in the Indian Ocean (an area 1.7 times the size of the British Isles). Unfortunately, global changes in ocean levels, temperature, and acidity do not respect the boundaries of marine sanctuaries. Construction of artificial reefs has been accomplished by dumping tires, concrete blocks, or other structures into coastal waters. But will we respond quickly enough to save a substantial portion of the world's coral reefs? If the rate of forest destruction is any indication, short-term commercial interests will continue to dominate until very few reefs are left. Then we envision a world where humans substitute fish for a diet that includes algae, jellyfish, and marine invertebrates, and the three-dimensional magic of a coral reef becomes a nice but distant memory, much like old-growth forests have become on land.

8.3.4 Forests

Globally, deforestation has resulted in emission of greenhouse gases as carbon locked in wood decomposes or is burned and the displaced forests no longer absorb carbon dioxide. Estimations for the next four years of tropical deforestation suggest emissions greater than those produced by aviation from 1903 to 2033. Tropical islands have a small proportion of the world's forests but halting deforestation there is still important in addressing this problem. Locally, forests provide islands with many benefits. Evaporation from forests helps promote rainfall on upper slopes, without which island aquifers would shrink. Forests also cool the air through evaporation of water from their leaves, counteracting global warming. Erosion control provided by healthy forests reduces soil loss from watersheds and sediment damage to coastal fisheries. Although lowland forests on most of the island groups have been replaced by agriculture and human settlements, some healthy forests remain in the critical upland watersheds. Protecting these will provide many direct and indirect benefits for islanders and island ecosystems.

8.3.5 Tourism and the carbon cost of travel

Tourism is now the mainstay of many island economies. Since the 1950s, tourism has grown because of the expansion of airplane travel (see Section 7.5.1). In some instances (Canary Islands, Jamaica, Hawai'i), tourism has become the dominant enterprise, replacing a series of failed agricultural industries. Unfortunately for the tourist industry and all those who enjoy flying, airplanes are major sources of greenhouse gases. One comparison suggests that a 5-hour jet flight contributes as much carbon to the atmosphere per person as driving a large vehicle for 1 year. Several approaches are being taken to address these concerns. Lighter planes that fly high up into the thinner atmosphere use less fuel. However, these may only be stopgap measures. What is needed is a radical new approach to tourism that might immediately include increased use of boats, or, unfortunately for islands, "stay-vacations" where people do not go far from home. Perhaps in the future, realistic cyber holidays will substitute for the real thing.

8.4 RESPONSES

It is reasonable to feel discouraged in the face of the plethora of environmental challenges that island ecosystems face. We highlighted only several challenges above (but see Chapters 6 and 7 for more details). In this section, we offer encouragement by recounting several ways that islanders are responding to environmental challenges.

8.4.1 Local production and consumption

The boom and bust economic cycles of agricultural products including sugar cane, pineapples, and coffee on islands, such as the Canary Islands, Jamaica, Puerto Rico, and Hawai'i, have exhausted natural resources and disrupted local employment. The loss of major export industries provides an opportunity for islands to recalibrate their economies. Tourism and both urban and suburban sprawl often appropriate agricultural land, permanently reducing the agricultural base for islands. Even when new residential or business developments are touted as green and job-promoting they still result in habitat loss for wildlife and removal of soil. Loss of local production of a wide variety of foods means more reliance on food imports. Tokyo, for example gets wheat from Australia, corn from the USA, and soybeans from Brazil. Fortunately, there are nascent movements to diversify crop production both to reduce disruptive cycles in exports and to reestablish local production. Local agricultural production and consumption reduces transportation costs and emission of greenhouse gases. Farmers' markets (see Fig. 6.22) and restaurants that serve locally grown products promote a small but important renaissance of living off the available land. However, most island economies are too dependent on imports. Hawai'i, for example, has only about two weeks of food and fuel on hand in case an emergency (such as a tsunami) were to disrupt local ports and make deliveries unreliable or impossible. Island nations therefore have a long way to go before they have reduced their imports to a level where local economies can be considered sustainable. Mainland nations face the same dilemma as do island nations of an over-reliance on imported food and fuel, although the causes and solutions to the dilemma may be less obvious than they are on islands where imports and exports are clearly measurable.

8.4.2 Restoration

Modern societies are facing challenges from loss of native communities that previously resupplied damaged areas, from increased human use of the land, and from novel impacts created by humans. Restoration can occur on land or in the water. We noted some efforts to restore coral reefs in Section 8.3.3. On land, reforestation has direct benefits for watersheds and is a widely used approach when land is too steep to use for agriculture. The benefits of slope stability and soil conservation provided by plantations of non-native tree species that are periodically harvested has to be evaluated against the benefits of promoting the use of native species, especially when native species grow more slowly and are less desirable for forestry than non-natives. In New Zealand, for example, radiata pine plantations cover large portions of former pasture land (see Section 7.3.2). Modern reforestation efforts provide promise by immediately reducing soil erosion and gradually improving water conservation. However, pressures mount for wood products and when not locally grown, these are imported.

8.4.3 Living with invasive species

Humans have accelerated the dispersal of so many organisms that the plants and animals of the world are now mixed in novel ways. Despite quarantines that reduce transport of non-native organisms to islands such as New Zealand or Hawai'i, the homogenization of the world's plants and animals continues apace. Fortunately, there is a growing awareness of the problem. Sometimes stands are taken against particularly onerous invasives that affect high profile, native species (see Section 7.6). However, successful delay of the spread of non-native species is rare. These novel communities may not be ideal from the perspective of preserving native plants and animals, but are an inevitable part of the future. We envision a future of altered communities where the line between native and non-native is blurred. Already, the term naturalized is used for non-native invasive species that have been around long enough to be considered a part of the community. We expect that the focus of conservation efforts will shift from issues of biodiversity to assuring that ecosystems provide adequate services (such as clean air and water) for humans.

8.4.4 Urban futures

Island populations reflect global trends of increasing urbanization. Because island ecosystems and the economies based on them are so vulnerable to habitat destruction, loss of native species, global temperature increases, sea-level increases, and increased fuel scarcity, island population centers have strong motivation to develop innovative approaches to these problems. Urban agriculture that uses rooftops, patios, walls, and parks can decrease reliance on distant food sources and lower heating and cooling costs. Europe has a long tradition of urban agriculture and citizens of London produce 14% of their own food. Wider development of modern transportation systems such as Japan's bullet trains can reduce emissions from vehicles. Retrofitting urban buildings with improved insulation and other energy efficient measures, as underway in London and Tokyo, greatly reduces energy needs. Restricting vehicles in inner cities such as done in London, Dublin, and Auckland, and promoting bicycles and buses is another approach to reduce greenhouse gases and make urban environments more people-friendly. In Japan, railroad stations sometimes provide multi-tiered bicycle parking lots because of substantial bicycle use by commuters. Finally, ecotourism has the potential to reduce environmental impacts of tourists. Cities that embrace green technology reduce costs and dependence on imports and contribute globally to lower emissions of greenhouse gases.

8.5.1 Application to other islands

The nine island groups that we selected based on where we have worked display such variability in physical and environmental characteristics that they can be considered representative of many other islands. Therefore, we can make some generalizations about islands. We now revisit our initial themes of isolation, finiteness, vulnerability, and evolutionary experiments (see Chapter 1) in the context of future challenges for islands.

The initial biological and cultural isolation of islands appears to have been greatly reduced with widespread connections now a reality through shipping, aviation, and telecommunication. This isolation was instrumental in the evolutionary pathways that island organisms took, but was lost when human colonization began. Near-shore islands, such as Japan and the British Isles, were colonized first and remote islands, such as Hawai'i and New Zealand, were colonized much later. European expansionism reduced island isolation during the 1700s and 1800s and the advent of jet travel in the 1950s completed the loss of isolation. Yet, island ecosystems remain relatively more isolated biologically and culturally than mainland ecosystems.

Islands have not lost their finite amount of space. Islands grow through the gradual processes of volcanism and geological uplift and gradually shrink through subsidence and erosion. Coastal areas are also now shrinking from rising sea levels. Inputs of nutrients come from newly exposed mineral soils when landslides occur and from volcanic ash deposits. Air-borne particles also provide critical dust-borne nutrients (to the Canary Islands and the Caribbean from Africa, or to Japan and Hawai'i from China) that can alleviate very long-term losses of soil nutrients. But for most biological processes, there are no new inputs of resources on islands. Instead, terrestrial ecosystems on islands are increasingly linked to the more extensive resource base available on mainlands. Marine ecosystems surrounding islands are less finite than terrestrial ones because of potential links to colonizing and migrating organisms. Major alterations of marine ecosystems come from reef building or sediment deposition from the land.

Vulnerability of island ecosystems has increased in direct proportion to the loss of isolation. Organisms that evolved in isolation have often been poor competitors when faced with mainland organisms that evolved to compete successfully against a much wider array of species. Small population size, small home ranges, no escape routes, natural disasters, and intensive human impacts have combined to eradicate many native island organisms with unknown consequences for future ecosystem functions. These same factors impact island plants and animals today as climate change introduces new challenges.

Evolutionary processes on islands have contributed substantially to our understanding of how species change. Now we can again look to islands to understand the impacts of non-native invasive species and the development of new mixes of native and non-native organisms into novel communities. These changes are occurring in the context of reduced genetic diversity on islands from reduced population sizes, prior extinctions, new forms of anthropogenic disturbances, and a steady influx of new species.

8.5.2 Practical lessons

The fate of island ecosystems rests in the hands of humans. If we choose, we can safeguard island ecosystems in a number of ways. First, we can do a better job of filtering what gets to islands. Second, we can reduce population pressures by raising living standards, which helps to reduce family size. Third, we can address the reduction of greenhouse gas emissions. Fourth, economic diversification helps stabilize island economies and reduces environmental impacts. Fifth, coral reefs could be better protected through expansion of sanctuaries and stronger controls over fishing, pollution, and acidification. Sixth, agriculture could make bigger strides toward producing for local markets. Seventh, undesirable species could be more vigorously targeted for removal as soon as they are identified. Finally, urban buildings and transportation could "go green" faster with immediate savings of fuel. There are hurdles to overcome, such as learning more about how humans impact the functioning of ecosystems and how best to restore damaged ecosystems. However, the biggest hurdles are political and social. Despite the knowledge that most of these actions would have immediate benefits to island ecosystems, too often the concern has been about maximizing perceived short-term benefits. Modern transport enables the wealthier inhabitants of islands to come and go frequently. Both Hawaiians and Jamaicans employ the term "off island" for residents who are temporarily abroad. For those abroad, "off island" is also a state of mind: it means that facing an island's problems can be staved off, at least in the short term. How much longer cheap air travel will permit this state of mind remains to be seen. Conversely, the rapid rise of modern communication means that it is only by choice that a person who can afford access to the Internet remains ignorant of life on distant islands, including the economic and environmental issues that these islands face.

8.6 SUMMARY

Many island ecosystems are in disarray. High dependency on imported fuel and food, increasing economic instability triggered by over-dependence on exports to volatile world markets, loss of many native species, and threats of large resource disruptions from global climate change leave islands with a questionable future. Humans have identified most of the problems facing island ecosystems and we are the cause of most of them. Ironically, environmental activism is often seen as counterproductive to human interests. Nothing could be further from the truth. Healthy ecosystems are essential to humans. Instead of taking ecosystem services for granted it is critical that we begin to value clean air, clean water, fertile soil, and biodiversity as much as we now value jobs and money. Island ecosystems can and do degrade as seen on each of the nine island groups we have discussed. Which ecosystem will be next? Instead of risking further loss of ecosystem services, it is urgent that we tap into the creative solutions that are at hand. Ideally, where ecosystems are only partially damaged we can alleviate the stresses (such as over-fishing) and allow them to naturally recover. Where systems have degenerated beyond thresholds of irreversibility (such as from soil loss) we can sometimes restore them. Where systems are permanently lost (through paving of agricultural areas or extinction of species) we must learn to live within the new order. Increasing efficiency of agriculture, using rooftops rather than fields, consuming and wasting fewer resources, and myriad other solutions can be applied. Islands have a future if we apply ourselves collectively to the wise management of the ecosystems and their resources upon which lives depend.

SELECTED READING

Brown, L.R. (2009). *Plan B 4.0: Mobilizing to Save Civilization*. New York: Norton.
Cohen, J.E. (1995). *How Many People Can the Earth Support?* New York: Norton.
Register, R. (2006). *Ecocities: Rebuilding Cities in Balance with Nature: Revised Edition*. Gabriola Island, BC: New Society Publishers.

Glossary

Abiotic: Pertaining to non-biological factors such as wind, temperature, or erosion

Active volcano: A volcano that has erupted in the last 10 000 years

Aeolian: Wind-dispersed, as for seeds and spores

Alien species: See invasive species

Anthropogenic disturbance: A disturbance caused by humans

Anthropogenic island: An island created by humans

Archipelago: A chain or cluster of islands

Atoll: An island of coral that encircles a lagoon

Bammy cake: A small cake made from cassava root

Barometric pressure: Atmospheric pressure or force per unit area exerted on a surface from the mass of air above it

Barrier island: An elongated, coastal landform that is just offshore and parallel to it; often composed of sand

Biodiversity: The number of species (richness) combined with the proportional distribution of species (evenness); commonly used to refer just to richness

Biome: A geographical region within which there is a similar vegetation and climate (such as a tropical forest)

Biotic: Pertaining to biological factors

Bronze Age: The era of human history in Asia and Europe that ended the Neolithic period, when natural outcroppings of copper were smelted into alloys using tin or arsenic to cast bronze used for agricultural implements, cutting implements, and weapons

Carrying capacity: The maximum number of individuals that a given ecosystem can support

Community: A group of species interacting at the same place at the same time

Continental fragment: An island that has broken off from a continental shelf of which it was once a part

Continental island: An island that is part of a continental shelf

Coral: Colonial animals that secrete calcium carbonate that hardens into coral reefs

Crannóg: An artificial (anthropogenic) island in a river or river mouth used for living or storage

Crustal plate: A rigid layer of the Earth's crust that moves slowly

Diatom: A group of algae with silica in their cell walls, a major component of phytoplankton

Disturbance: A relatively sudden disruption that results in loss of biomass, structure, or function. A disturbance can occur naturally or be human-made (anthropogenic) and has a particular frequency (return interval), intensity (force), extent (area), and severity (damage caused)

Ecological footprint: A measure of the resources needed to sustain a person, town, or other entity

Ecosystem: The sum of all organisms within a well-defined area, the physical environment, and the interactions between them

Ecosystem function: A process that defines the workings of an ecosystem, such as nutrient cycling or water retention

Ecosystem services: An ecosystem function that is useful to humans, such as removing carbon dioxide through photosynthesis

Endemic: A species that is only found within a given region

Estuary: A partly enclosed river mouth

Evapotranspiration: Sum of evaporation and transpiration from plant surfaces to the atmosphere

Exotic species: See invasive species

GDP: Gross domestic product

Geological fault: A discontinuity in a volume of rock, across which there has been significant displacement

Gene pool: The complete set of genes in a species or group of individuals

Genus: A taxonomic group containing one or more species

Geyser: A spring that explosively releases steam

Gondwana: A supercontinent located in the southern hemisphere that comprised the modern continents of Africa, South America, Australia, Antarctica, and many other continental fragments, including Arabia, India, Madagascar, New Zealand, and New Caledonia

Guano: Bird dung

Habitat: The area used by a particular organism or species

Hectopascal: A measure of barometric pressure (force per unit area), one hundred pascals

Hotspot: In geology: a zone of upwelling of hot magma; in biology: an area of unusual biological diversity

Interdunal swale: The low area between two dunes; often a wetland

Invasive species: A species that has moved into territory that it has not previously occupied; can include range extensions of native or non-native species (often called aliens or weeds if undesirable and exotics if desirable)

Iron Age: The era of human history in Asia, Europe, and Africa, characterized by the introduction of iron metallurgy, and which followed the Bronze Age. In Europe and East Asia it began about 2800 to 3000 years ago. Agricultural implements, cutting implements, and weapons of this era were made of iron or steel

Island: Land that is surrounded by water and smaller than a continent; islands can be part of a continental shelf (continental) or independent (oceanic or atoll)

Isostatic rebound: Rise of land formerly depressed by a massive weight, such as ice sheets

Karst: Irregular limestone terrain often perforated by caves and sinkholes

Kettle hole: Depression left in a floodplain by a melted iceberg

Land bridge: Land connection between two land masses

Loess: A fine, unconsolidated, wind-blown sediment

Magma: Molten rock found beneath the surface of the Earth

Makatea island: An uplifted limestone island

Mammal: A group of animals that feeds their young from mammary glands

Mantle plume: An upwelling of abnormally hot rock within the Earth's mantle

Mesolithic: The era of human history in western and northern Europe characterized by technologies associated with the transition from hunter-gatherer societies to settled agriculture. In northwestern Europe, this period occurred about 5000 to 12 000 years ago

Moor: Low-lying vegetation growing on upland, acidic soils

NATO: North Atlantic Treaty Organization

Natural disturbance: A disturbance uninfluenced by humans

Natural selection: The process by which traits that make it more likely for an organism to survive and reproduce become more common over time; a key mechanism of evolution

Neolithic: The era of human history that is usually associated with the beginnings of agriculture, especially cultivation of starch and grain crops, pottery, and permanent settlements. This era was characterized by use of stone implements, which were supplanted by metal tools in Asia, Europe, and Africa. Neolithic cultures continued until the nineteenth century in many parts of the world

Niche: The ecological role or space occupied by a species

Non-native species: See invasive species

Obsidian: Naturally occurring volcanic glass

Oceanic island: An island formed independently of any continent

Orographic precipitation: Precipitation caused by air rising due to topographic variation

Pacific Ring of Fire: An arc of intense seismic and volcanic activity circling the Pacific Ocean

Paleolithic: The era of human history that begins with the development of the first recognizable stone tools from about 2.6 million years ago, in the Rift Valley of Africa, until about 5000 to 15 000 years ago in various parts of the world

Phreatic eruption: An explosive eruption caused when rising magma encounters water

Primary succession: Succession on substrates with little or no initial organic matter

Polynesia: A group of over 1000 islands in the central and southern Pacific Ocean encompassed by Hawai'i, Easter Island, and New Zealand whose inhabitants share a common culture

Pyroclastic flow: Fast-moving currents of hot gas and rock that flow from an erupting volcano

Restoration: Effort to re-establish aspects of a damaged ecosystem

Rofabard: A remnant grass-covered soil patch in an eroded landscape

Sandur: Glacial outwash plain

Seamount: A mountain arising from the ocean floor but not reaching the surface

Shield volcano: A volcano with gradually sloping flanks built by relatively fluid lava

Shifting agriculture: Temporary use of forest clearings for growing crops, followed by abandonment of the site

Shogunate: The government of the hereditary commander of the Japanese army (the shogun)

Slash-and-burn agriculture: Cutting woodlands for agriculture

Soil order: A classification of soils based on diagnostic horizons

Stratovolcano: An explosive volcano that typically forms a steep, cylindrical cone composed of multiple layers of eruptive material including ash and lava

Subduction: The movement of one tectonic plate under another at a convergent zone

Substrate: Surface of ground or rock

Succession: Replacement of plant and animal communities by new communities over time in a given location

Tectonic: That part of geology concerned with movements within the rocky surface of the Earth

Tephra: Unconsolidated volcanic material that is thrown into the air

Topography: Shape of the Earth's surface

Tundra: A vegetation type of cold regions with low shrubs, sedges, and grasses

Ultramafic substrate: A surface of volcanic origin with a low silica content and a high content of magnesium and iron

Index

Note: bold page numbers indicate an illustration

'aina, 279
'elepaio, 223
'ō'ō, 140
'ohia, 120
acacias, 146
aeolian, 60
Aeonium, 138
Africa, 88, 163, 269
agouti, 162
agriculture, 3, 182, 183, 242–51,
    290
  centralization, 250
  ecological effects, 206
  feudal, 189, 190
  improvements, 305
  industrial, 10
  intensification, 243
  local, 302
  loss of land, 294
  mechanization, 244, 250
  monocultures, 243
  Neolithic, 10, 163, 183, **184**
  organic, 245
  pollution, 250
  production, 190
  slash-and-burn, 161, 173, 194, 195,
    248
  sustainable, 249
  technology, 190, 236
  urban, 303
Ainu, 200
air travel, 267, 271, 304, 305
  environmental costs, 301
albatrosses, 143, 201
alders, 155, 193
alien species. *See* invasive species
alluvial processes, 59
alpine plants, 196
aluminum, 255
Anguilla, 123

animals
  domestic, 147
  feral, 147
  Polynesia, 165
anthropogenic island. *See* island:
    anthropogenic
apples, 155
aquaculture, 10, 242
aquatic ecosystems, 14
aquifers, 299, 300
  artesian, 252
Arawak, 160
Archaic people, 160
ash, 155
Atlantic Ocean, 14, 29
atolls, 25, 130
aurochs, 157, 186
Australia, 121, 123, 140, 146
Azores, 136, 225

Baby Boom, 238
bamboo, 34, 265
bananas, 204, **205**, 207, 211, 247
Banks, Joseph, 214
barley, 163, 176, 178, 183, 192
barrier island. *See* island: barrier
basalt, 251
bats, 124, 167, 167
bauxite, 249, 251, **253**, 255
beaches, 83, 271, **273**
beans, 158, 192, 250
bears, 157, 158, 177, 186
beeches, 157
Berbers, 162
Bering Sea, 5
berries, 155
biodiversity. *See* diversity
biogeography, 214
biomes, 14
birches, 155, 177, 184

birds, 11, 38, 110, 130, 236
  colonization, 121, 123, 132, 133
  flightless, 140, 147, 167, 171
  Iceland, **42**
  Jamaica, **37**
  land-sea connections, 143
  New Zealand, 144, 174, 174, 213, 216
  nutrient inputs, 145
  Surtsey, 28, 118
  Tonga, 165, 167, 167
Bismarck Archipelago, 164, 166
bison, 157
blackbirds, 220
boars, 125, 158, 189
boas, 123
boats, 156, 163, 268
bombs, 236, 256
boobies, 143
Brachyglottis, 143
breadfruit, 209–10
British Isles, **238**, **246**, **261**, **262**, *See* England, Ireland, Scotland, Wales
  agriculture, 183–92
  coal, 252
  colonization by humans, 152–55
  Cornwall, **186**, **191**
  Coventry, 256
  dunes, 86
  geography, 40
  geology, 28–29
  Hadrian's Wall, **188**
  Hebrides, 28
  Liverpool, 256
  London, 29, 204, 256, 261, 266, 294
  Lundy, 286
  Manchester, 261
  map, **41**
  Norman invasion, 188
  Orkney Islands, **184**
  Outer Hebrides, 286
  Pennines, 52
  Roman invasion, 186
  Roman occupation, 186
  Shetland Islands, 155, 184
  sport hunting, 276–77
  Thames River, 266
  urbanization, 192, 261
  Yorkshire, **191**
bronze, 159
Bronze Age, 185, **186**
browsing, 133, 141
brushtail possums, **12**, 220, 278
bubonic plague, 175
buckwheat, 192
bulbuls, 229
buntings, 164
burrows, 144, **145**

cabbage trees, 217
cactuses, 204
canals, **262**
Canary Islands, **51**, **133**, **136**, **161**, **205**, *See* El Hierro, Fuerteventura, Gran Canaria, La Gomera, La Palma, Lanzarote, Tenerife
  colonization by humans, 164
  Cumbre Vieja, 70
  droughts, 107
  economic development, 202–07, 269
  European contact, 162
  export crops, 203
  forests, 202, 203
  Garajonay National Park, **81**, **136**
  geography, 41
  geology, 25
  landslides, **81**
  Las Palmas, 206
  map, **42**
  Pico del Teide, **51**, 131
  plants, 138
  Spanish conquest, 203
  tourism, 263, 268–70
  trading post, 203
  urbanization, 262–63
  volcanoes, 70
canistel, 160
canoes, 160, 162, 164, 166, 168, 173
carbon, 300
carbon dioxide, 300
Caribbean, 207, 280
Caribbean Plate, 29, 30
Caribbean Sea, 14, 88, 91
carmine dye, 204
carrying capacity, 6
Carson, Rachel, 239
cassavas, 162, 207, 208, 247
casuarinas, 146
cats, 135, 147, 163, 175, 213, 219, 229, 285
cattle, 175, 177, 207, 211, 225, 243, 278
caves, 171
Central America, 122, 160
cereals, 183
ceremonial ball court, **161**
chaffinches, 220
Chagos Islands, 300
chamois, 276
charcoal, 163, 169, 177, 185, 211
chestnuts, 187
chickens, 165, 207
Chile, 102
China, 93, 116, 198
citrus, 249
cliffs, **22**, 25, 131
climate
  measurements, 56–57

climate change, 3, 296–300
    coral reefs, 299–300
    forests, 300
    sea level and acidity, 298–99
    temperature, 297
    tourism and carbon cost of travel,
        301
climates, 50–57
coal, 236, 262
    power station, **246**
coastal property, 299
coastline, 23
cochineal scale insects, 204
cocoa, 211, 249
coco-de-mer, 116
coconuts, 229, 247, 249
cod, 241
Cod Wars, 241
coffee, 210, 211, **212**, 222, 243, 248,
    249, 301
Columbus, Christopher, 207
competition, 133
conservation, 10–11, 279–83
consumption, 10
continental fragment, 22, *See*
        island: continental
        fragment
continental island, 21, *See* island:
        continental
continental shelf, 22, 28
Cook, James, 145, 213, 214, 276
copper, 251
coqui frog, 278
corals, 21, 25, 240, *See* reefs
cormorants, 143, 144
cotton, 208, 261
crabs, 137
cranes, 189
    conservation, 282
    Japan, 282
crayfish, 240
cress, 145
crustal plates, 21, 25
Cuba, 30, 123, 159, 160, 161,
        269
culture, 267
Cyanea, **139**
cycads, 160
Cyclone Waka, **95**
cyclones
    animals, 94
    characteristics, 87–94
    damage, 94
    ecological effects, 94–96
    frequency, 88
    intensity, 89
    Kamikaze, 93
    plants, 96
    tracks, **86**
cypresses, 217

dairy, 246, 248, 250
dams, **255**
Darwin, Charles, 117, 132, 136, 214
DDT, 257
deer
    British Isles, 155, 189
    Hawai'i, 276
    Japan, 157, 197, 200, 276
    New Zealand, 276
deforestation, 3, 180, 232, 283, 287
    British Isles, 183–92
    Canary Islands, 163, 203, 206
    Hawai'i, 171
    Jamaica, 280
    Japan, 193, 195, 197, 256
    New Zealand, 173, 216
    Puerto Rico, 212
dengue, 297
diatomaceous earth, 251
diseases, 265, 297
dispersal
    land-bridges, 121
    long-distance, 6, 119, 123, 127, 168
    over sea, 118
    Surtsey, 118
    wind, 128
disturbance
    animal activities, 110–11
    anthropogenic, 7, 296, 305
    characteristics, 64–66
    cyclones, 87–96
    definition, 64
    droughts, 106–07
    earthquakes, 72–79
    ecosystem responses, 293
    erosion, 79–83
    exposure risk, 112
    extent, 65
    fires, 108–10
    floods, 96–101
    frequency, 65
    humans, 111
    intensity, 65
    interactions, **113**
    land building, 83–87
    natural, 7, 64, 66
    novel, 7
    regime, 65
    severity, 65
    tsunami, 102–05
    volcanoes, 67–72
disturbance regime, **113**
diversity
    genetic, 305
    habitat, 8
    New Zealand, 282
dogs, 147, 162, 163, 165, 173, 175,
        213
dolphins, 158
domestication, 155

Dominican Republic, 123
Drosophila, 120
droughts
    characteristics, 106–07
    ecological effects, 107
    plants, 107
ducks, 140
dunes, 84
    birds, 87
    coastal, 87
    plants, 87
    stabilization, 87
dwarfism, 140, 142
dynamite, 240, 251

eagles, 142
Earl of Bedford, 190
earthquakes, 3, **75**, **76**
    characteristics, 72–77
    ecological effects, 78–79
    magnitude, 73
    Mercalli intensity scale, 74
    moment magnitude scale, 74
    Richter scale, 72
earthworms, 176
Echium, 138
ecological footprint, 10
economic theory, 294
ecosystem
    function, 283
ecosystem services, 239, 306
ecotourism, 268, 274, 303
Ehrlich, Paul, 239
El Hierro, Canary Islands, 25, 162,
    203, 269
El Niño, 106, 107
elephants, 125
Eleutherodactylus, **137**, **138**
elevation, 56
elevational gradients, 131
Elingamita, 124
elms, 157, 184
emigration, 266
Emperor Chain, 26
endemic species, 131
endemism, 129, 130, 147, 252
energy
    alternative, 297
    renewable, 263
England, British Isles, **2**, 28, 39, 52,
    59, **191**
    cyclone, **90**
English Channel, 153, 183
environmental degradation, 232
environmental limits, 289
environmental movements, 238, 291
eradication, 285
erosion, 3, 8, 57, 59, **80**, 177, 250,
    302, 304
    British Isles, 28, 186

Canary Islands, 42, 70, 204, 262
Caribbean, 30
    characteristics, 79–82
    coastal, **84**
    ecological effects, 83
    habitat destruction, 10
    Hawai‘i, 45, 222, 225, 257
    Iceland, 179, 230
    Jamaica, 82, 232, 248
    Japan, 48, 197, 201, **202**, 236, 245, 257
    New Zealand, 30, 217, 245
    overgrazing, 81
    Puerto Rico, 82
    shoreline, 7, 17
    Surtsey, 28
    Tonga, 46
    volcanoes, **39**
    wind, 171
Ethiopia, 113
eucalypts, 146, 217
Euphorbia, **133**
Eurasian Plate, 33
evapotranspiration, 56
evolution, 7–8, 128–34, 214, 304, 305
    biological, 7
    biological features, 132–34
    cultural, 8
    island size and topography, 129–31
    isolation, 129
    time, 128–29
evolutionary divergence, 4
exotics. *See* invasive species
export crops, 232, 296
    boom and bust cycles, 301
exports, 247
extinctions, 146–49, 150, 171, 232,
    283
extractive industries, 251–56

famines, 198
farming, **205**
fences, 285, 286
ferns, 158, 174, 191, 213
ferrets, 220
fertilizers, 195, 236, 243, 244, 247
Fiji, 164
finches, 164, 217, 223
finiteness, 4–6, 21, 304
fire tree, 225, 278
fires, 147, 283, 297
    Blue Mountains, 248–49
    British Isles, 184
    Caribbean, 160
    characteristics, 108–09
    ecological effects, 109–10
    Hawai‘i, 225
    Jamaica, **109**
    New Zealand, 173, 216
    plants, 110
    Polynesia, 167

firewood, 194, 195, 201
firs, 221
fish, 155, 198
fish catches, 240
fish consumption, 241
fish ponds, **170**, 240
fish populations
  collapse, 242
fisheries, 213, 216, 236, 240
fishing, 3, **169**, 231, **242**, 272, 299
  exclusion zone, 241, 242
  rights, 241
  sustainability, 241
fjords, 41, 48
floodplains, 101, 186
floods, 100
  coastal, 98
  debris dams, 98
  ecological effects, 101
  flash, 99
  glacial outbursts, 97, 101
  rivers, 99
flycatchers, 121
food webs
  aquatic, 240
forestry, 246
forests, 153, 297, 300,
    See deforestation
  British Isles, 155, 185, 186
  Canary Islands, 163
  Hawai'i, 169
  Iceland, 177, 179
  Jamaica, 162
  Japan, 236
  koa, 223
  New Zealand, 173
  Polynesia, 213
  Puerto Rico, 161
  secondary, 250
  Tonga, 165, 167, 167
fossils, 152, 157, 169, 283, 285
founding populations, 133, 134
foxes, 145, 177, 201
frankincense, 116
frogs, 127, **137**, **138**
Fuerteventura, Canary Islands, 25,
    43, 162, 164, 203, 269
funerary monument, **186**

Galápagos Islands, 3, 117
gallinules, 140
gannets, 143
gene pools, 132
genetic diversity, 286
genetic engineering, 243
geography, 35–49
geology, 21–35
George Tupou I, King, 215, 228
giant roc, 116

gigantism, 140, 141, 161
ginger, 208, 211, 249
glacial advances, 125, 156, 157
glacial retreats, 153, 157
glaciers, 27, 125, 153
goats, 229, 278, 285
  British Isles, 184
  Canary Islands, 163, 202
  Hawai'i, 224, 258
  Iceland, 175
  New Zealand, 124, 148, 221
Gobi Desert, 114
gold, 251
Gondwana, 28, 127, 146
gorse, 245
gourds, 158, 168, 192
grains, 158, 163
Gran Canaria, Canary Islands, 25, 43,
    138, 163, 203, 269
grasses, 183, 215, 278, 279
grasslands, 189, 243
grayling, 220
grazing, 110, 147
  British Isles, 184, 189
  Japan, **202**
  moa, 141
  New Zealand, 133
  restoration, 284
greenhouse gases, 300, 305
greenhouses, **298**
Greenland, 175
greenstone, 251
greywacke, 59
Gross Domestic Product, **237**, **244**, 249
ground water, 254
growth
  limits, 5, 6, 294
  zero, 294
Guadeloupe, 123
Guanches, 163, 163, 202, 243
guano, 116, 118, 143, 144
guavas, 162
Gulf Stream, 54
gulls, 118, 143
gypsum, 251

habitats
  destruction, 8
Haiti, 75, 79
hares, 153
harriers, 217
Hawai'i, **5**, **122**, **134**, **166**, **170**, **170**,
    273, **287**, *See* Island of Hawai'i,
    Kaho'olawe, Kaua'i, Lāna'i,
    Maui, Moloka'i, Ni'ihau, O'ahu
  annexation by the United States, 222
  biomes, 8
  bird diseases, 224
  birds, 7, 11

colonization by humans, 168–72
cyclones, 91
dispersal to, 120
diversity, 120
earthquakes, 75
economic development, 221–26
erosion, **80**
extinctions, 172
geography, 45
geology, 25–27
Haleakalā, 56, 69
Halemaʻumaʻu Crater, **26**
Hanaʻuma Bay, 274
Hawaiʻi Volcanoes National Park, 101
Hilo, 102, 272
Honolulu, **18**, 103
introduced plants, 224
invasive ants, 13
Kīlauea Volcano, **22**, 25, 26, 54, 70
Kona, 271
Loihi, 25, 45
map, **45**
Mauna Kea, 45, 55, 121, **122**, 131
Mauna Loa, 45, 55, 67, 69, 108, 297
Papahānaumokuakea Marine
      National Monument, 299
Pearl Harbor, 256
sport hunting, 275–76
tourism, 274
tsunami, 103
Waikīkī, 271, 272, **273**
Hawaiian goose, 132
hazels, 185, 192
heaths, 177
Hebe, 134, **135**
hemp, 192
Henry III, King, 189
herbivores, 7
herbivory, 110, 149, 233
herbs, 143
herring, 241
Hispaniola, 29, 123, 159, 160
Hokkaido, Japan, 48, 157, 158, 200
   animals, 276, 282
   disturbance, 69, 93
   history, 125, 156, 157
Homo erectus, 152
Homo heidelbergensis, 152
honeycreepers, 139
Honshu, Japan, 48, 69, 82, 93, 102,
      125, 156, 158
Hooker, Joseph, 214
hornbeams, 163
horses, 175, **176**, 193
hot springs, 27, 119
hotels, **270**
hotspots, 27
housing, **261**
hula, 272

human ancestors, 152, 153
human colonization
   consequences, 180
human impacts, 8–13
human populations, 182
hunter–gatherers, 155, 182, 183, 231
   Japan, 192
Hurricane Gilbert, **91**, 248
Hurricane Ivan, 93, **95**
hurricanes, 3
hydropower, 252
   environmental damage, 255
   Iceland, 255
   New Zealand, 254

Iceland, **27**, **42**, **70**, **176**, **176**, **178**,
      242, **298**
   climate, **16**
   colonization by humans, 175–79
   Danish rule, 230
   dunes, 85
   earthquakes, 75
   ecomomic development, 229–31
   Eyjafjallajökull Volcano, 67
   fishing, 241
   geography, 40–41
   geology, 27–28
   glacial outburst floods, **84**
   glaciers, 97
   Heimaey, 69, **70**
   Laki Volcano, 69
   map, **42**
   Mt. Hekla, 103
   Reykjavík, 88, 231
   Skeiðarársandur, 85, **86**
   sport hunting, 275
   Surtsey, 28, 71, **119**, 128, 274
   Thingvellir, 27
   transplanted agriculture, 176
   Vatnajökull ice field, 40, **42**
   volcanoes, 69
iguanas, 123
immigration, 266
imports, 302
indigo, 208
Indo-Australian Plate, 24
Industrial Revolution, 192, 259
insects, 176
invasional meltdown, 226
invasive species, 7, 10, 277–79, 291,
      302–03
   dispersal, 277–78
   eradication, 12
   escaped pets, 278
   management, 11–13
invasives, 268
Ireland, British Isles, 39, 59, 79, 153,
      155, 190, 262
   dunes, 87

Irish elk, 147
Irish Sea, 153
Iron Age, 186
iron tools, 193
island
    anthropogenic, 33–35
    atolls, 39
    barrier, 33
    continental, 28–29
    continental fragment, 29–33
    corals, 39
    culture, 2
    limestone, 39
    makatea, 39
    number, 37
    oceanic, 22
    population pressure, 296
    resources, 182
    size, 37
    stereotypes, 1–3
    volcanic, 39
island biogeography, 117
island ecosystems
    environmental challenges, 301
    future, 306
    resource use, 295
Island of Hawai'i, Hawai'i, 25, **26**, 45,
    225
    birds, 223
    climate, 55
    invasive plants, 225, 250
    invasive species, 12
    lava, **22**
    military base, 257
    silverswords, 121, **122**
    soils, 57
    tourism, 271
    vog, 54
isolation, 4, 21, 304

Jamaica, **9**, **37**, **44**, **48**, **53**, **75**, **91**,
    95, **132**, **137**, **212**, **253**, **260**,
    270
    agriculture, 247–48
    bauxite, 252–53
    birds, 270
    Blue Mountains, 43, **44**, **48**, **52**, 61,
        92, **109**, 135, **137**, 248
    Cinchona Botanic Gardens, 248
    coastal resorts, 270
    colonization by humans, 159–62
    cyclones, 91, 92
    deforestation, **18**
    earthquakes, 75
    economic development, 207–12
    English conquest, 208
    export crops, 208
    ferns, 270
    fires, **109**
    geography, 43–44
    geology, 29–30
    Kingston, **91**, **95**, 212
    Kingston Harbor, 75, 85
    landslides, 82
    map, **44**
    Morant Cays, 38
    Nassau Mountains, **132**
    orchids, 270
    rainfall, 82
    Spanish conquest, 207
    tourism, 270–71
Japan, **32**, **39**, **127**, **202**, **255**, **264**,
    295, See Honshu, Hokkaido,
        Kyushu, Shikoku
    agriculture, 244–45
    Arasaki, 282
    climate, 157
    colonization by humans, 156–59
    cyclones, 93
    Daibutsu-den, **195**
    earthquakes, 76
    fishing, 242
    forest management, 192–202
    geography, 48–49
    geology, 31–33
    Hiroshima, 236, 256
    Japanese Alps, 196
    Kagoshima, **104**
    Kansai Airport, 34
    Kikai, 69
    Kobe, 77
    landslides, 82
    map, **49**
    Matsushima Bay, 82
    militarization, 201
    Mt. Fuji, 48
    Nagasaki, 236, 256
    Nagoya, 294
    Nara, **194**, **195**
    Odaiba, 33
    Ogasawara Islands, 126
    Okinawa, 126, 156, 257, 286
    rice cultivation, 192–202
    Ryukyu Islands, 69, 82, 93, 196
    Sakurajima Volcano, 103, **104**
    sport hunting, 276–77
    Tokyo, 77, 126, 198, 256, 266
    Tokyo Bay, 33
    tsunami, 103
    Unzen Volcano, 103
    vegetation, 157
    volcanoes, 69, **104**
    Yakushima, 196, **199**, **200**
Java, 112
jelly fish, 240
jökulhlaup. See floods: glacial
    outburst
Jōmon culture, 156, 158, 158, 192

Kahoʻolawe, Hawaiʻi, 25, 171, 256, 286
Kamchatka, 112
Kamehameha III, King, 221
karst, 43, 44, 59, 252
Kauaʻi, Hawaiʻi, 25, 45, **134**, 223
    erosion, **80**
    extinctions, 140, 141
    landslides, 82
    rainfall, 54
    rare plants, **126**
    tourism, 271
kauri, 216
kava, 247
kelp, 297
kettlehole, **86**
Korea, 93, 156, 199
Krakatau Volcano, 6, 103
Kuril Islands, 159
Kyushu, Japan, 69, 103, **104**, 125, 126, 156, 282

La Gomera, Canary Islands, **81**, **136**, **161**, 203, 269
La Palma, Canary Islands, 25, 74, 90, 164, 203, 269
Labrador, 175
lakes, 98
    craters, 98
lamb, 230, 243
Lānaʻi, Hawaiʻi, 25, 104, 171, 222, 223, 271
land bridges, 22, 28–29, 37, 125, 153, 155, 156
land building
    characteristics, 83–87
    ecological effects, 87
land reform, 244
land use, **169**
landfills, 264
landslides, 3, **58**, 79
Lanzarote, Canary Islands, 25, 43, 162, 203, 269
Lapita, 165, 166
    pottery, 164, 166
lapwings, 217
laurel, 163
lava, **22**, 26, 57, 58
    types, 71
leisure, 267–77
leopards, 157
Leopold, Aldo, 238
lichens, 191
limes, 207, 251
limestone, 21, **24**, **53**, 59, 60, 131
limestone cliffs, 44
Little Ice Age, 178
lizards, 164, 167
lobelias, 141

local production and comsumption, 301–02
locusts, 163
loess, 60
logging, 197
London, Jack, 271
Lord Howe Island, 128
lupines, 287, **288**

macadamia nuts, 251, 272
macroclimate, 56
Madeira Islands, 136, 225
mahogany, 208
maize, 162, 208
makatea, 130
malaria, 210, 224, 297
malasadas, 272
mammals, 132, 133, 147, 157, 232, 284
mammoths, 153, 157
manatees, 161
mangos, 212, 272
mangroves, **170**
Māori, 173, 237
    crops, 8, 213, 215
    European contact, 148, 213, 215
    fires, 172
    Māori-British wars, 216
    sacred mountains, 279
    trade, 174
marble, 251
marijuana, 250
marine organisms, 4
marine sanctuaries, 299
Marquesas Islands, 121, 168
martens, 191
Maui, Hawaiʻi, 25, 56, 103, 271
    silverswords, 121
Mediterranean climate, 50, 108
Melanesia, 166
Melville, Herman, 271
mesoclimate, 56
Mesolithic, 155, 155
metal technology, 231
metalwork, 185
mice, 163, 164, 175, **188**, 203
microclimate, 56
micronations, 34
mid-Atlantic ridge, 72
middens, 172
Middle Ages, 175
migration routes, **154**
military
    bases, 290
military activities, 256–59
    environmental benefits, 257
    environmental consequences, 256
military construction
    United States bases, 257

military training
    protests, 258
milk, 185
millet, 244
mines
    reclamation, 252
mining, 198
moa, 7, 103, 111, 123, 141, 174
moa-nalo, 141, 171
Moloka'i, Hawai'i, 25, 45, **281**
    dunes, 85
    fish ponds, 240
    human population densities, 171
    landslides, 82
    tourism, 272, 273
Mongolia, 282
Mongols, 93
mongooses, 210, 223
monkeys, 123, 161, **199**
    Japanese macaque, 197
    Puerto Rico, 277
monks, 175
moorlands, 186, 189
moose, 157, 276
mosquitoes, 224
mouflon, 276
Mount Pinatubo, 65

national parks, 239, 280
National Trust, 280
native animals, 285
native plants, 285
native species, 3, 13
natural resources, 235
    extraction, 236
natural selection, 140
nēnē, **134**
New Caledonia, 128, 164
New Zealand, **12**, **31**, **47**, **71**, **127**,
    132, **135**, **145**, **149**, **284**,
    289, *See* North Island, South
    Island
    agriculture, 245–47
    animal introductions, 220
    Auckland Islands, 175
    Avoca River, 77
    Banks Peninsula, 216, 217
    birds, 7
    Campbell Island, 144, 285
    Canterbury Plains, 217
    Chatham Islands, 124, 175
    Christchurch, 217
    colonization by humans, 172–75
    conservation, 13
    cyclones, 90
    dispersal to, 123
    diversity, 123
    dunes, 85, **86**
    earthquakes, **76**, 77
    economic development, 215–21

European settlement, 174
floods, **99**
forestry, 246
geography, 46–48
geology, 30
glaciers, **97**
hydropower, 254–56
introducted plant species, 221
Kapiti Island, 13
Kermadec Islands, 175
Lake Manapouri, 254
Lake Taupo, 104, 254
landscape, 217
landslides, 82
map, **47**
Maud Glacier, **97**
Milford Track, 274
moa, 7
Mokihinui River, 256
Mokohinau Islands, 129
Mount Cook National Park, **97**
Ngauruhoe, 69
Ruamāhuanui, 144, **145**
Ruapehu, 69, 98, 104
Southern Alps, **47**, 52, 77, 146,
    217
sport hunting, 276
Stewart Island, 278
Taupo, 71
Three Kings Islands, 124, 128, 129,
    149
Tongariro National Park, 279
volcanoes, **111**
Wellington, 90
whaling, 214
Whanganui River, **99**
White Island, 47, 69, **71**,
    111
Ni'ihau, Hawai'i, 25
nitrogen, 57, 226, 287
nitrogen-fixing bacteria, 226
non-native species. *See* invasive
    species
Norfolk Island, 128
Norse agriculture, 177
North Africa, 163
North American Plate, 29, 33
North Atlantic, 88
North Atlantic Ocean, 118
North Atlantic Plate, 30
North Island, New Zealand, 47, 71,
    86, 129, 148
North Sea, 29, 98, 155
Norwegians, 175
novel biological communities,
    278–79
novel ecosystems, 278, 291
nutrients
    inputs, 304
nuts, 155

O'ahu, Hawai'i, 13, **18**, 25, **40**
  birds, 223
  cliffs, **40**, 45
  forests, 169
  rainfall, 81
  tourism, 271, 272
oaks, 135, 157, 163
obsidian, 156, 163, 174, 251
oceanic island. *See* islands: oceanic
oceanicity, 54
oil, 263
Olearia, 143
Operation Bootstrap, 263
opossums, 162
orange roughy, 240
oranges, 207, 247
otters, 215, 221
overpopulation, 295
owls, 172, 220
oxen, 193
oystercatchers, 206
oysters, 240

Pacific Ocean, 3, 14, 102, 103
Pacific Plate, 24, 25, 33
Pacific Ring of Fire, 67, 102
Paleolithic, 153
palila, 223
palms, 116, 132, 167, 169, **170**
Pangaea, 29
papayas, 162, 272
Papua New Guinea, 164
parrots, 274
peaches, 192
peanuts, 162
pears, 155
penguins, 143, 144
Pennantia, **149**, 148
pesticides, 244, 247
petrels, 144, **145**, 210
pharmaceutical companies,
    254
Philippine Sea Plate, 33
phosphorus, 57
phytoplankton, 241
pigeons, 124, 167, 203
pigs, 135, 175, 201
  Europe, 155, 163, 207
  New Zealand, 221, 276
  Polynesia, 165, 213, 276
pimentos, 211
pineapples, 222, 243, 271, 301
pines, **154**, 155, 155, 204, 217, 221,
    246, 302
pioneer plants, 83
piopio, 220
plantains, 207
plants
  buried, 71
  defenses, 149

Polynesia, 165
  wind-dispersed, 71
plate tectonics, 26, 35
Pleistocene, 50, 125
ploughs, 185
poisoning, 285
polecats, 191
pollen, 169, 171, 177, 184
pollination, 121, 126, 167
pollution, 262, 263
  air, 191
Polynesia, 165, 167, 173, 247
  economic development, 213–29
  transported landscape, 169
Polynesian migrations, 240
poplars, 217
population growth, **191**, 267
populations
  carrying capacity, 295
  wealth, 294
Portlandia, 131, **132**
potatoes, 214, 250
pottery, 157, 159, 160, 160, 164
pouākai, 142
poverty, 263, 271
precipitation, 50, 54, 61, 99
predation, 133, 142, 164
predator pressure, 132
predators, 7, 285
Pritchardia, 169, **170**
Puerto Rico, **17**, **58**, **161**
  agricultural abandonment, 284
  aquifers, 253–54
  Caribbean National Forest, 101, 274
  colonization by humans, 159–62
  Culebra, 258
  cyclones, 91
  ecological footprints, 10
  economic development, 207–12
  geography, 43
  geology, 29–30
  landslides, **80**, 82
  Luquillo Mountains, 135
  map, **43**
  San Juan, 35, 212, **218**, 263, 277
  Spanish conquest, 207
  United States invasion, 212
  urbanization, 10, 263–64
  Vieques, 257, 258
Puerto Rico Trench, 75
puffins, 40
  Iceland, **42**
pūkeko, 140, 142, 217
pumpkins, 196
pyrethrum, 257

quails, 164

rabbits, 7, 189, 219
radiation, **138**

rails, 131, 140, 210, 215
rain shadows, **51**, 52, 131, 173, 216
rainfall, 99, 101, 106, 131, 171
    intensity, 100
rainforests, **17**, 131, 135, **136**, 173, 248
rats, 219, 223, 229
    agents of extinctions, 232
    Canary Islands, 164, 203
    human dispersal, 147
    New Zealand, 173
    Norway, 7, 175, 192, 214
    Pacific, **166**, 169, 171
    Polynesia, 165
    removal, 285, 286
    ship, 187, 207, 210
    types, **188**
recreation, 267
recycling, 263
reefs, 25, 59, 160, 240, 268, 271, 297,
    302, 305
    loss, 299
reforestation, 232, 302
    Japan, 201
    Puerto Rico, 212
reindeer, 5, **6**, 153, 275
remittance cultures, 259–66
remittances, 10, 282
resources
    imported, 290
restoration, 13, 283–88, 302
    goals, 283, 285
    soil, 245
revegetation, **284**, **288**
rice, 192, **194**, 197, 250
rivers, 294, **295**
rock phosphate, 117
rodents, 161, **188**
rofabard, 60
root crops, 160
rowans, 177
rum, 250
rural, 259
    percent of population, 237
Russia, 156, 282

Sahara Desert, 60, 84, 114, 269
salt marshes, 190
Samoa, 102, 165, 168
sand, 83
sand plains, 85
sandalwood, 116, 221
sandwort, **119**
sanicles, 120
scenery, 267
scientific voyages, 214
Scotland, British Isles, **22**, 28, 29, 39,
    52, 59, 153, **156**, 190
    dunes, 87
scurvy, 145
sea cows, 240

sea level rise, 298
sea levels, 3
sea lions, 144
Sea of Japan, 31, 52, 125
sea urchins, 240
seafood, 160
seals, 144, 158, 214, 215, 236
seedlings, 167
seeds, 128, 129, 166, 167, 285
    dispersal, 127
sewage, 262, 265
sharks, 240
shearwaters, 143, 144
sheep
    British Isles, 184, 189
    Canary Islands, 163, 202
    Caribbean, 207
    Hawai'i, 258
    hunting, 276
    Iceland, **176**, 229, 230
    New Zealand, 217
shellfish, 155, 158, 158, 160, 163, 165
shifting cultivation, 211, 216
Shikoku, Japan, 125, 156
ships, **270**
shrews, 157
shrublands, 155, 173, 177
Siberia, 156
silver, 251
silverswords, 120, **122**, 276
silviculture, 242
slavery, 203, 207, 208, 209, 210, 211, 248
sloths, 161
smallpox, 194
snail jails, 286
snails, 172, 201, 214, 277, 286, **287**
snakes, 125, 278
snapper, 240
snorkeling, 268, 274
soil, 57–58, 147, 179, 186
    animals, 110
    erosion, 57, **178**
    fertility, 59, 195
    formation, 57, 61
    geological substrate, 58–60
Soil Conservation Service
    Iceland, 287
soil formation
    biological influence, 61–62
    climatic influence, 61
    topographical influence, 61
Sonchus, 138
South America, 88, 160, 160, 269
South American Plate, 29
South Atlantic, 88
South Island, New Zealand, 47, 52,
    76, 77, **97**, **99**, **135**, 173, 217
    landslide, 82
South Pacific, 90
Spain, 269

Spanish conquest
  Canary Islands, 163
  Jamaica, 162
  Puerto Rico, 162
spatial scales, 4
spear fishing, 268
species composition, 283
species richness, 135–36
spiders, 57
spines, 141
sponges, 240
sport hunting, 275–77
spruces, 157
squashes, 243, 247, 251
squirrels, 192
St. Matthew Island, 5, **6**
standard of living, 296
starch crops, 213
starlings, 124, 229
State of the environment in 236, 250
State of the environment in 251–56,
  290
Stevenson, Robert Louis, 271
stoats, 220, 285
stone circles, **184**
stone terraces, **205**
stone wall, **191**
storm surges, 83, 93, 98
stowaways, 278
streams, 254
subsistence fishing
  Tonga, 241
suburbs, **264**
succession, 65, 83
sugar, 243, 248, 250, 280
sugar cane, **9**, 60, 196, 203, 208, 222,
  232, 251, 301
sugi
  Japan, 196
sulphur, 252
swallows, 217
sweet potatoes, 168, **170**, 196, 247
sycamores, 191

Tahiti, 209
tahr, 276
Taíno, **161**, 162, 207, 208
Taiwan, 93, 126, 164
takahē, 140
tapirs, 135
taro, **161**, 163, 165, 243, 247
tarweeds, 120
Tasman Sea, 103, 123
Tasman, Abel, 148
Te Heuheu Tukino IV, 280
tea, 196
teak, 212
technology
  ecological consequences, 235
  information, 236

tectonic activity, 4
temperate climate, 55
temperature, 50, 55–56, 61
temperature gradients, 131
Tenerife, Canary Islands, 25, 42, **133**,
  164, 203, 269, 270
  disturbances, 81, 90
  history, 164
tephra, 59
terns, 143, 144
terrestrial ecosystems, 14
textiles, 261
thorns, 141
time series, 57
tobacco, 162, 207, 208, 249
todies, 135
Tokara Strait, 126, **127**, 159
Tonga, **24**, **95**, **169**, **170**,
  See Tongatapu, Vava'u
  'Eua, 25, 131, **170**
  agriculture, 247
  birds, 7
  colonization by humans, 164–68
  cyclones, 90
  earthquakes, 75
  economic development, 228
  fishing, 241
  Fonuafo'ou, 24, 46
  geography, 46
  geology, 24–25
  Ha'apai Group, 25
  Hunga Tonga-Hunga Ha'apai, 69
  Kao, 24, 46
  land allocation, 228
  land sales, 215
  Late Island, 24
  Late'iki, 24
  map, **46**
  Niuatoputapu, 102
  Nuku'alofa, 24
  subsistence agriculture, 228
  Tofua, 24, 209
  volcanoes, 69
Tonga Trench, 24
Tongatapu, Tonga, 24, 25, 69, 104,
  164, 164
topography, 8, 51–52
toucans, 135
tourism, 10, 197, 291
  air travel, 237, 271
  attractions, 267
  Hawai'i, 272
  in 237, 250
  island attractions, 275
  shoreline development, 273
  sustainable, 271, 273
  two-edged sword, 267
trade winds, 51, 88
trapping, 285
trees, 163, 167, 169

tropical climate, 55
trout, 220
Tsugaru Strait, 125, **127**
tsunami, 3, **5**
    animals, 105
    characteristics, 102–05
    ecological effects, 105–06
    plants, 105
    submarine landslide, 104
tuatara, **132**, 136, 144
tundra, 153, 184
tunnel, **289**
turtles, 165, 201, 208, 240
tuyas, 40
Twain, Mark, 271

urban, 235
    agriculture, 303
    buildings, 303, 305
    ecology, 265
    futures, 303
    heat island, 56
    percent of population, 237, 259–60,
        291
    planning, 265
    pollution, 265
    sprawl, 261, 264, 301
    sustainability, 265
    vehicles, 303
urbanization, 10, 233, 259–66, 291
    Japan, 199
    loss of species, 265

vanilla, 229, 247
Vanuatu, 164
Vava'u, Tonga, 25, **95**, **169**
vegetable crops, **200**
Venezuela, 269
Vietnam, 165
vog, 54
volcanic ash, 60, 71
volcanoes, 3
    characteristics, 67–70
    ecological effects, 70–72
    intensity, 68

voles, 157
vulnerability, 6–7, 304

Wales, British Isles, 52, 59
Wallace, Alfred, 117, 214
walled fields, 185
walls, **191**
walruses, 177, 178
water tables, 287, 297
waterfowl, 155
wattlebirds, 220
wealth, 267–77
weasels, 220
weedy species. *See* invasive species,
    See invasive species
weevils, 119
Western Sahara, 162
wētā, **127**, 134
wetlands, 190
whales, 158, 214, 240, 241, 242,
    274
whaling, 221
wheat, 163
white-eyes, 226
wildcats, 191
Wilkesia, **126**
willows, 177, 217
wind farms, 281
wines, 204, 243
wolves, 142, 189, 190, 200
wooden buildings, **195**
wool, 185, 230, 243
World War I, 204, 217
World War II, 235, 236, 258,
    290
    environmental damage, 236
    Japan, 244, 256, 257, 266
worms, 241
wrens, 127, 164, 219, 220

yams, 165, 209, 247
Yayoi culture, 193, 194
Yucatán Peninsula, 160

zinc, 252